KT-526-116

➲ Acknowledgments

I would like to thank all of the people who supplied me with information for use in this book: all of the help desk vendors, industry groups, trade associations and consulting groups. There are too many of you to mention.

I would like to thank The Gartner Group and Dataquest for all their help and extend special thanks to Bill Keyworth, Bill Kirwin and Tammy Kirk of the Gartner Group, and Tom Sweeny and Erin Collier at Dataquest for helping me find answers to the questions posed to them and for sharing a generous portion of their research with me. I would also like to thank the Help Desk Institute for sharing research published in their Help Desk Customer Support Practices Report.

I would like to thank all of the help desk vendors who took the time to offer their opinions and answer questions posed to them, many of whom shared Requests For Proposals and other valuable information. Through the years, the time they've spent showing and explaining to me how their products work have helped give me a clear understanding of the technology.

I would like to thank the consulting groups, committees, consortiums and individuals quoted throughout this book for providing their input on a variety of issues. They have made themselves available to me for questions, providing valuable input for several of my help desk columns in Call Center Magazine and for this book.

Special thanks also goes to Robbie Alterio for her terrific layout of this book and to Christine Kern for getting this book to you.

This is my first book. I encourage all readers to share their thoughts and opinions, noting what you liked, did not like, what you found to be lacking, etc. Your suggestions can help me make changes and improvements for any future updates to this book.

Send me an e-mail at MLenz@mcimail.com.

CONTENTS

► CONTENTS ◄

➲ Foreword: Why I Love Help Desks

by Harry Newton

O ne night our dishwasher broke. My wife said, "repair it." "It's 2 AM," I said. "It will wait." "It won't," she replied, "We have dirty plates." I grunted.

I called KitchenAid's 800 number (800-422-1230), hoping they'd be asleep. They weren't. I explained what was wrong. They gave me two repair options. I tried one, while they waited. Bingo. It worked. I'd fixed the dishwasher. I thanked KitchenAid profusely. My wife loved me again. We went to bed.

I'll never buy any other manufacturer's dishwasher again, I thought. That should have been the story's happy ending. It wasn't.

When Mary asked me to write a Foreword to her book on help centers, I called KitchenAid's Help Desk to ask some simple questions. They have totally screwed it up. They have replaced their nice, helpful people with a really stupid interactive voice system. It keeps asking you to push a button. "Want to buy a parts manual? Push 1." "Want to find your closest repair shop? Push 2." etc. There are more buttons to push than I can count. I didn't want to buy anything. I didn't want one of the options. I wanted to speak to someone. Eventually, after 10 minutes on hold, I got to speak to a real live human being.

I asked, "What happened." A nice lady told me "We don't do technical support any more." I asked why. She said "it was upper management's decision." Was it a good decision, I asked. She said NO. "We liked that line," she said, "It was called 'Do It Yourself' and it really helped our customers." I asked her why upper management had made the decision to eliminate it. She didn't know. I bet it was to save money. One of those good short-term decisions, but very dumb long-term decisions. I thought to myself, that's the last KitchenAid appliance I'll ever buy.

Fortunately, there are other companies — like most of KitchenAid's competitors — with more enlightened management. That management understands how important customers' questions are. They understand that the stuff we all sell today — from kitchen appliances to insurance policies, from overnight package delivery to software — is complex and raises questions in our customers' brains. Those questions might be as simple as "Where is my package?" Or as complex as "Will this cosmetic cause my allergies to go nuts?" No company today can afford to be without a Help Desk.

Every company needs a Help Desk for their customers and for their employees (who need help on medical, insurance, education, etc.) For three years, Mary Lenz, the author of this book, has written a superb monthly column on Help Desks in Call Center Magazine, the biggest call center magazine in the world.

Mary is the world's leading expert on Help Desks. I am privileged she asked me write this Foreword. Please do not judge this book by what I've written. I write from the heart. She writes from the brain. That's more useful.

P.S. Guess what? GE's Answer Center on 800-626-2000, option 4, will answer questions on their appliances at any time of the day or night, including 2:00 AM. Guess whose dishwasher I'm buying next? You're right. It's not KitchenAid. *©*

— Harry Newton New York, NY October, 1996 harrynewton@mcimail.com

(Harry Newton is the publisher of Call Center, Teleconnect, Imaging and Computer Telephony magazines.)

➲ INTRODUCTION

Simply defined, a help desk is an area in any organization that helps end-users find answers to questions or find solutions to problems. These end-users or "customers" may be people who have bought products or services from the organization they seek help from. Such help desks are often referred to as support centers or external help desks.

Other types of end-users or "customers" who seek assistance may be employees in an organization who need help with a situation that has arisen in their workplace. Such a situation is likely to be a problem using computer software, accessing the network, a printing problem, or some other kind of technical question. Many of the questions may be as simple as requiring the password to reset their workstations or as complex as CMOS error. The area set up to assist employees in a company is typically referred to as a help desk or internal help desk.

The help desk set-up in an internal organization can be as informal as visiting computer whiz John Smith down the hall to get help or as structured and well-executed as calling a dedicated number to reach a technician knowledgeable in the specific area an employee needs help in. We will talk about the problems with the informal underground support network (visiting John Smith) later in this book.

For the purposes of this book we will use the term help desk interchangeably when referring to both:

- a support center that takes calls from external customers and

- an internal support department that takes calls from internal employees.

Through this book we will explain the differences and similarities in both types of support. We will also look at why employees should be treated the same as paying customers when requiring assistance from the help desk.

We will explore how help desks came to be, why they continue to grow and why they should be considered one of the most important departments in any organization.

The sophistication of the technologies that are used in a help desk vary depending on the size of the help desk, the type of support required, budget and technical level of support reps and end-users. Client/server-based systems have undoubtedly made technology more accessible and cost efficient for the smaller desk. We will take an in-depth look at systems used in help desks and give you all the information you could have ever wanted to help you make the right decisions when looking for new technologies that will save your organization time and money.

Some of the technologies we now see have drastically changed the way the help desk solves problems. The Internet is a tool that can be a best friend or enemy to the help desk.

All of the information presented in this book comes from my research of the help desk market. Through the years I have spoken to countless help desk managers and talked with several industry groups to explore the issues concerning support and the many challenges facing help desks. I have met almost every vendor that makes problem resolution tools and evaluated almost all of their products.

While this book will not solve all the problems of the help desk, it will provide useful advice for choosing the right tools for automating and putting together useful practices within the help desk. This book with tell you about the help desks put together by successful companies. It will help you decide whether or not you should outsource certain help desk functions. It will tell you how to put together a Request for Proposal and give you examples of RFP's put together by companies looking for the best help desk system. This book will also tell you about processes being worked on today that will eliminate many of the problems the help desk faces in the future.

Most importantly just about everything discussed in book serves one goal — the goal every help desk manager should have. That goal? How to provide the best service to customers while cutting costs and increasing productivity. This is the only way the help desk can prosper and be looked upon as an asset to the entire corporation. ©

➲ Chapter 1:
The Dawn of the Help Desk

Why They Developed

When we look back to equipment used in the workplace in the 1970s and 1980s and compare it to equipment used today, the differences are dramatic. Back then, some organizations created IS Departments to support dumb terminals. The help desks that we see today typically support networks, software and hardware used by employees in that organization or customers who purchase items sold by organizations that were virtually nonexistent only a couple of decades ago.

In comparing these IS Departments to the typical help desk of today, it seems more like a century than a decade has past.

The number of PC users in the US has almost doubled compared to three years ago. According to Dataquest, there were 49.3 million PC's shipped in the US in 1993, 5.9 million in 1994, 71.8 million in 1995,and Dataquest is expecting shipments to reach 85.5 million by the end of 1996.

The proliferation of computer use in both the corporation and the home has bred a generation of non-technically oriented people who need assistance from specialists to use this equipment.

In the past, problems were less complex. Prior to the computer age, people used simplistic items in the workplace like ledger books, typewriters and the good old pen and paper. The most complex equipment used at home was probably a stereo or VCR.

Today that's all changed. The majority of today's corporate working population use computers at their jobs. These new PC users tend to not be technically inclined. While many home users purchase PCs to allow them to work from home, many also use PCs to run accounting programs or educational software and games for their children.

Computer sales will continue to flourish for quite some time. This means that the demand for technical support for the complex questions using computers and software generates will continue to increase.

In corporations it's much more than just the advent of PCs that have made support more complex. The way computers are used has created a complex environment. When PCs first emerged in offices, they were standalone workstations. The need for support existed, but the complex software of today and the local area network did not.

LANs and WANs make every workstation a part of the entire corporate enterprise. The need for network managers came into play as all the issues associated with running software and PCs on servers needed to be managed. In the smaller corporation a couple of dedicated people may share responsibilities for tracking inventory, corporate assets, supporting the network, change management and help desk functions.

PC ownership costs in the distributed computing environment have risen dramatically with the improvements in hardware and software costs being buried by tremendous increases in labor costs. In 1987 the Gartner Group conducted its first analysis of PC ownership costs for corporate computing environments. At $19,296 in 1987, the total cost of ownership of a DOS based PC today has grown to over $40,000, more than a 150% increase.

Tips for Buying Problem Resolution Tools

What it boils down to in the end is choosing the functionality that best fits your organization. Not every help desk package is applicable to every organization. These are some questions that can be answered through a careful review of product literature and a set of evaluation disks.

1. *Is the software easy to install, learn and use? If you have problems so early in the game, it seems reasonable to assume that you will have problems throughout the life cycle. As a rule, quality software is intuitive and therefore easy-to-learn and use. Speedy on-line (customizable) graphical systems with built-in help are the recommended environment.*

2. *Does the software conform to your current operating environment of choice? Choosing a help desk should not require the organization to revamp its environment. This means that the help desk should run in conjunction with the software environment the organization is now using including its network and database of choice. It comes as no surprise, therefore, that the majority of organizations desire a client/server help desk solution. This way, the organization can take advantage of its individual environment by dispersing the help desk engine to the server and the client interface to the PC, if desired. In addition, the help desk software of choice should permit the organization to add modules (i.e. expert system tools) when appropriate.*

3. *Does the help desk provide ODBC support? As in point two above, a robust help desk environment works "with" rather than "against" the current operational environment. From this perspective, ODBC support is crucial. ODBC, which is an acronym for Open Database Connectivity, is a Microsoft-driven specification (quickly becoming a standard) for an application program interface that enables applications to access multiple database management systems using Structured Query Language (SQL). ODBC permits maximum interoperability. In other words, the Help Desk's knowledge base should be equally at home on SQL Server as it is on Sybase, DB2, DB2/2, Oracle, Access, Paradox, Ingres or even Btrieve.*

4. *Does the help desk support simultaneous user access? Unless you're planning to have only one support analyst, it is important that the software you choose enable multiple analysts - even end-users - to access and update the knowledge base.*

5. *Are alternative access paths to the help desk supported? Not all users will use the telephone to notify the help desk of problems. A good product will enable end-users to e-mail problems. A good product will also enable the help desk analyst to e-mail, fax or page the problem resolution out.*

— Professional Help Desk

For example, Bill Kirwin of the Gartner Group cites that administrative costs have quadrupled and end-user operation costs have doubled. He says that although the strategic value of PC technology is increasing, the cost of technology may become prohibitive if enterprises do not take aggressive cost-reduction actions.

They also found that in 1987 the average PC was a stand alone machine running one or two applications. In 1994 the average PC was connected to a network and ran at least four applications.

Also, in 1987, a representative number of machines in the average enterprise was 600. In 1994, that number increased to 2,500.

Based on some of these results, Kirwin syas they developed Total Cost of Ownership (TCO) comparison models for major platforms.

Tips for Buying Problem Resolution Tools

Answer the following questions before you go looking and you should be on the right track:

Determine the type of support or problems that need to be addressed (these may involve more than networks).

Where is this happening?

Define specifically how many staff members, locations and/or devices require support.

What resources do you have to solve these problems?

Define the numbers, locations, expertise available in-house and otherwise.

What do you want the solution to do and how soon?

Define what needs to happen to solve the problem — prioritize these items, include any integration with other in-house systems and add time frames for staged and/or complete solution delivery (include resources as well as hardware/software and line issues).

Who can do this?

Match the solution needed with available resources and corporate problem resolution procedures.

What can we do to make this happen?

Given budgets and resources, determine the approach to take, deciding if you need to add staff, train staff, set-up in-house centralized support, distributed support, outsourced support, and so on.

Set up an evaluation.

Depending on the size and scope of your operation, answers to the above may be easily mapped out or they may require additional assistance. Whatever the case, once you can answer these questions you are ready to issue an RFP and/or set up to evaluate a number of different products that suggest they can help provide the solution you are looking for. The more precise you are about your needs and available resources, the more accurately you will be able to evaluate various products. Almost all commercial systems will miss something you want, so:

Be clear on what issues are most important for your particular operation. Know what legacy and/or database systems you have and will NOT change. This is key for compatibility issues and optimum operation of most systems out there. Know what features you are willing to wait for. Make sure that custom work or features are real before you buy. Don't forget training and future compatibility — a little training can go along way in getting things organized particularly for multi-user and distributed systems.

— STEPS: Tools Software, A Division of Cauchi Dennison & Associates Inc.

Using Windows 95, Windows 3.1 or IBM OS/2 platform has proven to cut costs. For example, the Gartner Group found that Windows 3.1 reduced the amount of formal and casual learning required by end-users by 44%. Kirwin suggests that enterprises operating in a Windows environment should plan to migrate to Windows 95 for reduced total cost of ownership and pay back in improved use of application features and improved user productivity.

Formal Technical Support

It may seem that the only real revenue a help desk generates is money that comes from support contracts. WordPerfect is an ideal example of a company that tried to provide free support and failed. The costs for such service cut too deep into profits. Today it is rare to find a company that does not charge for live technical support, after providing free support for the first 60 days after purchase.

Upper management has long viewed the help desk as a black hole draining their profits, and charging for support is often not enough reason for them to okay spending for automation or making other improvements. Management does not realize that a small investment in the help desk can go a long way.

A company that employs people who use computers and software may realize that when users have system problems they rely on other employees with more technical expertise for answers. What may not be so obvious are the costly repercussions of such actions.

When no formal help desk exists, or an overly simplistic, over-extended help desk exists that is ineffective at helping end-users, corporate employ-

Smooth Sailing Tips

Strive to become a profit center for your company by providing the highest possible level of customer service at the lowest possible cost. Review the level of service and support offered by external software companies noted for outstanding customer support and follow their lead.

Budget and allocate resources carefully based on the most accurate historical information and estimates of future needs possible. Remember to include new hardware and software implementations, number of new hires, new software releases and upgrades, etc. in your planning. Help Desk Managers often underestimate requirements for personnel and equipment.

Analyze support calls to determine trends and patterns. If you understand what kind of support calls you are getting and why, you may be able to reduce the number of calls through improved documentation, training, etc. These solutions are often easier, faster and more economical than simply increasing the number of support reps in response to increased call volume.

Increase the efficiency of your department with improved training, tracking, and reporting. Using automated customer support and defect tracking software to assist you in this task can transform the help desk into a profit center for your company. It facilitates training by giving all reps access to solutions provided by the most knowledgeable person on staff through a shared database. It enables you to track calls, improve internal and external communication, balance workloads, and assign tasks. Analyzing the data collected by the software assists you in demonstrating the effectiveness of your efforts, as well as pinpointing areas where improvements are needed.

— Repository Technologies

ees are forced to seek help from other employees. These people are not employed to handle support issues but are sought out because they are considered to have more technical expertise than other employees.

These employees have their own responsibilities, yet until the problem is solved, both the employee seeking help and the employee trying to supply solutions may remain unproductive. A sticky problem may even involve several employees putting their heads together to figure out a solution. Eventually this informal network will become overburdened with requests for support. Productivity is wasted since the same problems will be solved repeatedly among various groups. This is completely counter-productive.

According to the Gartner Group's Bill Kirwin, formal technical support programs are chronically under-funded. This shortage, which shifts support activities to the arena of end-user operations, is the greatest single factor in the escalation of total cost of ownership. For every $1 cut from the formal support team, $2 is spent for underground support.

I've been told by countless help desk managers that the biggest reason for not purchasing effective help desk software is that they have not been able to convince upper management to make the investment. This is a very sad situation since the return on investment when using an automated problem resolution system can occur within only one year.

Help desks need to convince upper management of the negative effect a help desk with poor service has on end-users. Since the help desk is often considered a low priority in the organization as a whole, funding for it is scarce. The help desk must promote their functions and demonstrate how spending a little money upfront to reorganize and effectively automate will save a small fortune down the road.

When end-users do not get the support they need the true cost of support will only increase. The best way to create a cost-efficient help desk is by being one step ahead — stopping the situation before it becomes a problem that will generate a call to the help desk.

Utopia/help desk features tab folder design, configurable tool bars, MDI and utilization of Windows standards for ease of screen and program navigation.

It's only recently that so much literature and vendor marketing has been touting ways to be "proactive." Yet being proactive is undoubtedly the key. There are now systems that can scan a local area network correcting a situation itself or alerting the right person before the situation turns into a problem that will generate calls to the help desk.

There are systems that can send e-mail when a printer is not working so employees are aware of the problem instead of needing to call the help desk and ask why. There are now systems that even automate the process of resetting a printer or terminal when the user forgets his or her password.

Being proactive results in call avoidance. The less calls to the help desk translates into less staff required to solve problems. And that translates into dollars.

Tips for Buying Problem Resolution Tools
Call Tracking and Management

1. Does the help desk provide automatic call tracking? When the telephone rings, the analyst should be provided with the highest level of automation. The knowledge base should enable the analyst to track a large amount of information about the client and his problem. Trackable information should include: name, client ID, department, phone, fax, address, analyst ID, date call opened, date call closed, priority level of call, subject of problem, resolution of problem, session notes, hardware environment, software environment. It would also be useful if the help desk provided the ability to create custom fields. Because calls come rapidly, the help desk software must be able to add new users, applications, etc. on the fly. Alternatively, the organization must be able to archive data and remove user names from the active user list without deleting associated problems.

2. Does the help desk enable the analyst to forward and/or escalate the call to another analyst or superior? The help desk analyst should have the ability to easily reassign the case to another analyst or superior at the same or higher level of priority. The software, however, should be able to retain the fact that the case has been reassigned, in effect logging the progress of the case every step of the way.

3. Are there flexible reporting facilities? Standard reporting is usually insufficient, therefore the ability to perform ad hoc reporting is essential. The ad hoc reports created by the organization should be storable and recallable. In addition, there should be an ability to do a quick print of a problem ticket.

4. Are there management facilities? Managers and authorized analysts should be provided with a graphical representation of status of all calls by analyst, group, call length, date or other specified criteria.

5. Does the help desk software provide the facility to track both hardware and software inventory? Organizations are finding that they must keep accurate records of their stock of hardware and software programs from both a financial as well as legal point of view.

6. Is the help desk software intelligent? Everything we've mentioned so far comes under the guise of call tracking and management software. While full of functionality, alone, this does not constitute a "smart" help desk. For help desk software to be fully functional, it should be intelligent. Call tracking and management functionality must be coupled with artificial intelligence techniques.

— Professional Help Desk

➲ Chapter 2:
Setting Up a Help Desk

Goals & Mission

Defining your goals is one of the first things you should do whether implementing a formal help desk for the first time or reevaluating an existing help desk.

Some of these goals should include:

- setting up a help desk today that will grow with the company for years to come.

- Investing in technologies that will fulfill your present needs and not be obsolete in five years.

- Staying cost efficient.

- Taking a proactive role in tracking expenses so you can show upper management how you are being cost effective.

- Keeping customers satisfied

- Keeping employees satisfied

While fulfilling all of these goals may sound like a tall order to fill, knowing what your goals are before you begin the project of re-engineering an existing help desk or starting up something new will help you stay focused. And once these goals are clearly defined, trying to reach them will encourage the help desk to prosper.

You should also consider writing a mission statement that's simple and easy-to-remember. The general mission of the help desk should be to function as a single point-of-contact for solving end-user problems. A mission statement should be a tool to empower your employees.

To come up with an appropriate mission statement, look at the products and services you produce and break them down into the business processes that support these products or services.

Some examples of what a mission statement may be: Customer service is a number one priority; A statement of service level goals; Always responding courteously on the phone.

Starting from Scratch

First, decide why you need a help desk. If you are considering starting a help desk it should be obvious that people who use your organization's resources or buy your company's products are in need of technical assistance beyond what your customer service or IS department can adequately provide.

Next, you must decide on the types of questions and problems that will be generated from your end-users. To anyone with an average level of technical expertise, the typical questions and problems users will encounter may seem obvious. But as any seasoned help desk manager can probably vouch, you would be surprised at the situations you don't expect.

You need to have a plan for addressing these calls. Will you offer a Web site so customers can post questions? You must also think about the percentage of calls you can eliminate by using call avoidance strategies.

Once you have decided that you need to set up some kind of area where customers (external users or internal employees) can call for assistance, you need to estimate the number of people that will require support in a given period of time (such as between 8 am and 6 pm, considering when call volume will be highest). This will help you figure out how you will need to staff. Here is a simple model you may want to use as a guide:

- Do I want customers to use online manuals, an intranet or the Internet to try and find an answer to their problem before calling the help desk? If so, you should poll end-users, asking about problems they've encountered in the past two months. Then take the top twenty problems and their solutions and post them online.

- For the next two weeks, ask end-users to go online to check the bulletin board you've set up when they have a problem to see if they find the solution. After two weeks find out how many people were successful in this approach, and if this approach served as a quick fix to their problem. Use the number of end-users who were not successful as an indicator for the number of calls you should expect in a two week period, if you continually post common problems and their solutions online.

Tips for Buying Problem Resolution Tools

Problem resolution tools should be fast and easy to use. Multiple screens, user interfaces, and steps make a problem resolution tool useless. By the time the poor help desk operator finishes clicking, typing, and navigating through the screens of some help desk applications the phone is on fire! In a perfect world several of the expert systems such as case based reasoning and the like would work well. Unfortunately no one other than large companies with relatively large budgets can spend the time to author cases and build/integrate knowledge bases. At a recent support conference full-text searching was voted most popular.

—Opis

This model will only work if you do not expect the number of new end-users to change drastically any time soon.

This model also won't work if you're planning to use a new software program or installing new computers corporate-wide. The same goes if you are releasing a new product. These three scenarios will surely bring an influx of end-users in need of support.

Keep in mind that this is a small scale approach for a company just getting their feet wet with providing support to a small number of end-users.

Once you've thought about call volume and the types of calls end-users are likely to generate, you can think about hiring staff.

Choosing the right people is critical to the success of your help desk. Based on the extent of the services you wish to provide here are some things to consider:

• Decide on the number of positions you want to fill.

• Decide on the technical expertise you want staff to have. Will first level support be entry level people with a general understanding of what it is they will support? How experienced should second level staff be? Is there a need for a third level support tier for the extremely challenging problems?

• What are your resources for training staff?

• What kind of call tracking/problem resolution system do you plan to use? A product that takes advantage of artificial intelligence technology will give you a higher resolution rate with less experienced support reps but there are higher costs associated with such products. (see chapter 3, sections on knowledge bases and artificial intelligence).

Reevaluating an Existing Help Desk

If you are considering changing the way you run your support operation, things must not be running perfectly. Maybe you're still using a pen and paper method to track calls (if so, you are desperately in need of a re-evaluation and some automation). Perhaps you've realized that it's time to scrap your home grown call tracking system. Or maybe upper management has told you to revaluate because costs are too high, or they think it's about time they had the cutting-edge support operation of their competitors.

Here are some questions to ask yourself and some preliminary re-evaluation steps to take:

• What exactly are you unhappy with?

• What does it cost to handle each support call? Figure out the number of calls you get in a year (that should take into account busy months, perhaps times when you've released a new product) and the average length of time of calls last (take into account the 30 minute calls and two minute calls to get an average).

- Call customers that have phoned for support in the past three weeks. Ask them if they remember how satisfied they were with the support they received.

- Look at staff. Is turnover high? Are reps happy handling the types of calls they're taking?

(See more about installing the right practices in Chapter 7: Managing your help desk.)

Controlling Calls and Costs

You might think that there is little you can do to prevent an influx of calls to your support center. That's partially true. Customers will always call your desk with problems your support reps need to troubleshoot. But you have control over the number of customers who call you. Options like electronic support, internal charge-back and automation tools help you control the number of customers who will call for support. Staffing adequately is also critical.

Electronic support reduces call volume. More and more people are signing up for Internet services everyday. Your end-users require only a modem to take advantage of the valuable troubleshooting information you can put online. You can put your knowledge base online to let users dial in and search the system the same way a tech would.

Most vendors of automated help desk products now offer links to the World Wide Web that let your end-users use the same troubleshooting techniques your techs would use to solve problems.

Tips for Buying Problem Resolution Tools

1. Analyze your current and future needs and create a list of criteria to narrow your evaluation. Factors to consider include: importing existing data, ease and speed of implementation, ease of use, user-friendly graphical interface, manufacturer's service and support.

2. Look for strong severity/priority tracking with complete incident audit trail and characteristics based linking of incidents, problems, and resolutions.

3. Look for full problem management and defect tracking facilities with strong inter-departmental communication mechanisms and task assignment, assign/reassign capabilities.

4. Look for flexible rules based escalation and alerts to ensure that customers receive rapid response and nothing "falls through the cracks".

5. Look for maintenance contract management, including expiration warnings and the ability to provide a variety of maintenance types: pay by time, pay by call, block of calls or incidents.

6. Look for enterprise-wide client / server architecture.

— Repository Technologies

Another way to decrease call volume is to give customers the option of logging problems online where technicians can check requests periodically and post solutions. The customer can later go back online to check the status of their problem.

"The impact is incredible," says Jeffrey Tarter, editor and publisher of Softletter newsletter (Watertown, MA). "It's like using automated teller machines. After a while customers will prefer this method."

Internal Charge-back

When developers are made responsible for calls that come into the support center, they are more likely to develop products without some of the quirks or hard-to-use features users have problems with. These situations generate the most calls.

"When you can track a call by product and charge it back to the product group there's pressure for the group to make the product more supportable and think about the complicated features," says Tarter.

Tarter doesn't see charging customers for support as solving any big problems since most calls tend to come in while the product is under warranty when support is free. "Customers don't typically abuse support," he says. "Although upper management sees the customer as the one with the problem — not knowing how to use the product — that's usually not the case. The incentive to develop an easy-to-use product should rely on the developer."

For example, Tarter says the support center may suggest that the developer fix an install routine to eliminate many problem calls. Often, the developer will argue that such a thing really isn't a problem. But once 50% of the calls that come are about this problem,and the developer is charged for the calls, he or she will have an incentive fix the problem. Then it's just a matter of sending out new disks rather than staffing to field all of these unnecessary problem calls.

Automating

Not enough emphasis can be placed on automating your support center. It is probably the biggest money saver — automating lets reps handle more calls in less time. They can use features within the system that help them find the right solutions. Help desk software can escalate open tickets and put all the information reps always had to look for in manuals right at their fingertips.

Smooth Sailing Tips

Provide help desk analysts and technicians with the same technologies being used by their customers. Many times the help desk is the last to get new versions of software or hardware which makes it difficult and stressful for them to support these technologies.

— McAfee

"One third to half of most calls are spent collecting routine information," says Tarter. "The beauty of automating is that it eliminates the need to ask about configuration, when the product was purchased, what else the customer has on their hard drive, and then looking through reference books. Technicians are turned on by solving problems and helping people. They don't want to spend time entering routing information."

Outsourcing

Outsourcing support functions has become a popular move. Rather than paying salaries, medical coverage, rent on extra space, and purchasing sometimes costly equipment, it is usually more cost-effective to outsource at least some of the functions of the help desk.

"Outsourcing is an enormous embarrassment to the support center," says Tarter. "Support centers are asking how can outsourcers afford to handle a call at $6 when it costs us $25 and customer satisfaction is just as good. It puts pressure on software publishers to explain why their support center is less efficient. That's why support is being outsourced left and right. If it's cheaper why not?"

We'll talk a lot more about electronic support, automating and outsourcing in upcoming chapters.

Predicting Call Volume

Knowing how to staff and manage queues in your help desk will let you decrease head count while handling calls more efficiently.

If you ran a call center for a catalog company, you would not staff agents from 9 a.m. to 5 p.m. That would be ludicrous. Experience would tell you to look at when the majority of calls came in and when there were lulls. Then

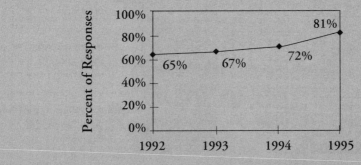

Call Volume on the Rise

For the fourth year in a row, HDI member support centers reported an increase in call volumes. Of those who answered the survey (1,040 centers responded), 81% of the support centers reported an increase incall volumes, up from 72% just a year ago. Source: The Help Desk Institute.

you would staff to meet call volume and service standards. The same concept applies to your support center.

"Tech support uses staffing to determine the number of people they need to handle peaks," says Tarter "Then they get excess capacity, because they don't take into account slow times. They are wasteful of people and it forces their costs to go through the roof."

Tarter contends that support centers staff for flat eight hour days,not paying attention to hours when they should staff up or down to meet call volumes. "Support centers may allocate a certain amount of time for callbacks, training and research, but they don't monitor peaks and valleys, says Tarter. "Some desks, like Novell and Intuit, are just learning. Once they staff properly there are enormous gains."

Tarter says most tech support people he comes across say they never really thought about scheduling part-time support techs. "One person even told me they did not have enough desks," he says. "Upper management doesn't think about these issues. They've never touched part-time labor or thought about queues. They hate support. They see it as a black hole they keep shoveling money into."

Tarter says another big problem is that support centers set up multiple queues for no reason. This is a big productivity deflator. "WordPerfect had a queue set up with a group of reps who dealt only with setting up tables in WordPerfect," says Tarter. "They would get a lot of calls, but techs also spent a lot of time staring at the walls. They figured out that if they created a pool of first level agents who could answer anything they could have only one queue. They almost doubled productivity."

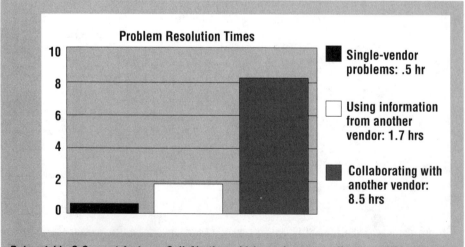

Datawatch's Q-Support features Call Alerting which sends an e-mail or page if outstanding calls come close to exceeding time limits for being solved.

It's common knowledge that the release of a new product or upgrade will generate more calls, so staffing up is a necessity. Take Microsoft for example. Windows 95 first went on sale at midnight one day in 1995. People were on line at computer stores across the country to make their purchases. Microsoft recorded their first support call at 12:09 a.m., just minutes after the product first went on sale.

You can just imagine what call volume was like in the following hours and for the next several weeks. But did Microsoft staff so that no caller would ever get a busy signal? Absolutely not. In any organization, if callers never have to wait for an agent or get a busy signal, it's a sign of over staffing.

If Microsoft staffed so that every caller was required to wait an average of twenty to thirty minutes before reaching live support, they could get away with it. Their reputation is so that someone buying Microsoft products would probably not say "I won't buy this because if I need support I will have to wait on hold for a long time."

In reality Microsoft has achieved monumental success by providing electronic support. A very high percentage of their customers use the Internet or bulletin boards rather than calling in for support.

Few companies can afford the luxuries of a company like Microsoft. If you are a small- or medium-sized software company, you're likely to face competition. You can't afford to leave callers hanging on for over twenty minutes on average. Yet it doesn't have to be expensive to staff based on call volume. You just need to staff in relation to call volume.

No one is suggesting staffing a help desk like a hotel reservation desk or a catalog company. These organizations make more sales the more calls they handle, unlike the help desk. Still when customers have to hold for long periods of time it is undoubtedly frustrating, especially when they are paying for support.

There are systems available that take the brain work out of staffing. These systems are known as call center management systems. Using such a system does not require a dedicated person to use spreadsheet programs to keep track of everyone's schedules. These systems take data pumped out

Smooth Sailing Tips

Give every member of your support team a non-support related task (such as newsletter writing, employee training, etc.) that makes use of their support knowledge in a proactive manner. Invest in an application that lets each member of your staff build his/her area of expertise into the application so that others can share in their knowledge. Provide either (or both) an on-line and electronic mail interface for your users/customers to log their own problems (and check on the status of past problems they've logged) in your database. This both reduces front-line call volume and lets your reps "get a feel" for a caller's problem prior to speaking with them.

— Applix

by phone switches and put it to smart use.

The right management software can make a help desk manager's job much easier by taking the guesswork out of what call volume will be and telling you how you need to staff. To learn more about the systems available, consult Call Center Magazine, 212-691-8215.

Here's just one example of a government help desk that's effectively using call center management software.

The Post Office Delivers

The Post Office's Minneapolis-based help desk has 33 agents that handle over 3,000 calls a week. These calls come from post offices around the country (including Puerto Rico and the Virgin Islands) that have software and hardware related questions.

"We have 10,000 internal customers who use distributed systems and mainframes," says Mary Conklin, Manager of ISS for the Minneapolis center. She says all data is collected at the offices and then uploaded to the mainframe in Minneapolis. More than 800,000 employees have time cards that get downloaded into the mainframe so paychecks can be issued every two weeks. The help desk supports all of these data transmissions.

"We currently use Infoman from IBM for problem solving but are planning to use Software Artistry's Expert Advisor so we can eliminate using manuals," says Conklin. "We want an expert system where we can find information in one database."

Prior to last October Conklin says abandonment rate was too high and they were using an old obsolete ACD and monitors. Supervisors had to manually pull numbers from the system. Now they use Northern Telecom's (Nortel) Meridian ACD and Telecorp Products' (Walled Lake, MI) ACD Performance Software, Agent Watch (now called Agent Window) and LED readerboards. "Now callers don't wait more than a minute and we aim for abandonment rate no higher than 3%," says Conklin. "We have more flexibility and I can get multiple information from the software.

She says agents are able to manage themselves better since they can view ACD stats either through their phone display or on the readerboards. Conklin prefers the readerboards because everyone can see statistics "at a glance."

Soon their five support centers will consolidate to three and call volume at Minneapolis will double. Then they will have 53 agents trying to solve problems and answers questions while maintaining service standards.

Structural Ergonomics

No matter how many people your help desk employs, quality in the environment — everything from adequate lighting to well-adjusted workstations

— affects your employees. That affects their productivity. It's imperative to design and maintain an employee friendly workplace.

Because reps perform many repetitive phone and keyboard tasks and are likely to spend all day (excluding breaks and lunch) at their desks, using ergonomic equipment is crucial. You'll get happier, healthier and more productive employees. In the long run your organization will save a bundle in time and money since you'll have lower turnover and better morale.

There are several ways to minimize work-related health problems:

- encourage people to take breaks;
- buy adjustable furniture;
- teach people how to adjust that furniture for comfort;
- and train people how to perform keystrokes with less strain.

Considering there are more employees suing now than ever before for repetitive stress injuries (reported incidents of RSIs are higher than ever, accounting for more than 60% of all occupational illnesses) there's no better time to offer courses in prevention and re-evaluate your center's set-up.

The following are some of the factors to consider when building or rehabilitating your help desk. Taken together, they'll help you create a productivity-enhancing workplace.

Workstation Design

Cluster workstations are popular in the support center because they are the most practical. Cluster workstations make the most of space, while still creating individual workspace for a computer and phone. Laura Sikorski, design consultant at Sikorski-Tuerpe & Associates (Centerport, NY), says cluster workstations are best for team building. "In a pod of four, senior reps can make calls while still keeping an ear out to listen to trainees," says Sikorski.

For reps who handle multiple tasks within their work areas, the Department of Labor and the Occupational Safety and Health Administration (OSHA) recommend L-shaped work surfaces because such an arrangement allows the employee to swivel the chair between tasks (e.g., writing, attending to a printer, in addition to using a PC.

"While everyone wants to be paperless, there's always reference material that reps shouldn't have to reach to get," says Sikorski. "Desks should be as clutter free as possible. The solution: putting a catalog rack at a 45 degree angle so material can be slipped in between and shared by two reps, says Sikorski.

She also recommends hanging a "hot file" (like an in/out rack) on the end of desks to help keep the work area clear.

Tab Products (Palo Alto, CA) makes the model cluster workstation. At the cluster's center core there's space for voice, data and electrical equipment, both within reach of each of the four users.

Each user can also decide on the amount of air flow they want. A built in air cleaner removes tiny particles from the air.

CenterCore (Exton, PA) also makes cluster workstations in different sizes (four pod, six pod, etc.) with a variety of storage options. Their Airflow 2000 has built-in air filters to purify and recirculate the air. This filtration system removes airborne particles and draws heat away from equipment.

Bramic Creative Business Products (Ontario, Canada) makes furniture specifically designed for people who sit in front of computers all day. They've designed a sit/stand computer workstation so agents can alternate sitting with standing while they work on the keyboard. A lift trolley brings the keyboard up to standing position.

Their furniture is designed to work with chairs from Grahl Industries (Coldwater, MI). These chairs have adjustable backs, seats, armrests and elbow supports. These elbow supports were designed to help protect against RSI and Carpal Tunnel Syndrome. Their "Duo-Back" or "Hugger" chair is split into two sections to support both sides of the users back without putting pressure on the spinal column.

Seating

The seat and backrest of every chair in your center should support a comfortable posture, permitting occasional variations in the sitting position. Chair height and backrest angle should be easily adjustable, says OSHA. They also recommend the chair height allow the entire sole of the foot to rest on the floor or a footrest with the back of the knee slightly higher than the seat of the chair. This allows the blood to circulate freely in the legs and feet.

Gere Picasso of consulting firm Engel Picasso Associates (Albuquerque, NM), stresses the importance of using ergonomic chairs, and even more important using them correctly.

"There is a learned helplessness syndrome," she says. "People need to realize that their posture is different at different times of the day and they need to adjust their chairs when they begin to feel uncomfortable. There's also a notion that one size fits all, when really very few people fit into the norm."

At two 400 employee support/call centers Picasso helped design for software manufacturer Intuit, she recommended chairs that were adjustable height wise, but also had adjustable arms that could expand to properly seat a heavier person and retract for a smaller person.

They also installed foot rests, monitor lifts and adjustable keyboards where the tilt and the pitch can be adjusted.

Thanks in part to their extensive program orienting employees to the workplace over the past year and half, Picasso says Intuit has reduced their insurance costs significantly.

"A company can help safeguard itself if they can demonstrate that everyone uses ergonomic furniture and are properly trained about ergonomics," says Picasso. "It makes the chances of liable less."

The Pronto Series of chairs from Girsberger (Smithfield, NC) follows the user's movement, locking into any desired position. They have styles available with adjustable height backrests and armrests. The seat and backrest can be adjusted independently of each other.

Lighting

Indirect lighting is best, says NIOSH researcher, Dr. Putz-Anderson. "Scatter screens help reduce glare," he says. "Parabolic diffusers, which are half-inch by half-inch squares, reduce glare compared to a typical screen used over fluorescent lights which cause scatter. Over shoulder lights should be avoided because they cause glare. It's best to use indirect lighting to minimize scatter. Parabolic diffusion screens are usually seen in newer office buildings."

He suggests replacing plastic panels which cover fluorescents with these parabolic diffuser screens. He also says lighting should be bright enough to not require supplementary light, but too bright is not good either.

OSHA recommends arranging workstations and lighting to avoid reflections on the screen or surrounding surfaces. There should not be any glare. When light reflects on a VDT screen or other reflective surface it causes a glare (a harsh, uncomfortably bright light).

To limit reflections from walls, OSHA recommends painting walls a medium to dark color, with a non-reflective finish.

Laura Sikorski says to avoid workstation surfaces that are too light or too dark. "If the color is too light the lights will be too reflective and cause a

Smooth Sailing Tips

Know processes and needs before changing entire systems.

Ensure the solution chosen not only meets current needs, but allows for support of evolving needs.

Make sure the solution chosen allows for integration with technologies relevant for the success of the system's operation.

Choose vendors wisely - make sure the product integrates with the appropriate databases and that the vendor shows commitment to expanding support as necessary.

— Remedy

glare," says Sikorski. "If too dark the light will be absorbed. She recommends something in between, such as light oak finish.

Natural Illuminating Technologies (Gaithersburg, MD) makes lighting designed to create sun-like illumination. The lights are color balanced and fit into existing fluorescent fixtures. They also have an ultraviolet filter for eye and skin safety.

Whatever products you choose for your help desk, make sure to examine them carefully or hire a consultant to help you out. Not everything on the market is all it's cracked up to be. "Over the past two or three months I've looked at 23 new add-on products," says Gere Picasso.

"With all the new stuff coming down the pipe you really need to look at what the product offers and the long and short term costs. Some I wouldn't give a nickel for effectiveness and durability, yet these products are often the most expensive."

Acoustics

Here are some no no's and recommendations from Laura Sikorski to give you the best sound quality in your center.

- "Do not paint ceiling tiles when refurbishing," says Sikorski. "This takes away their sound absorption." Rather than painting walls, she says wall vinyl is much better for sound absorption.

- If the center has very high ceilings use acoustical clouds to help absorb sound. Plants also help. "Not only are they good for putting oxygen into the air, they help absorb sound," she says.

- Avoid pictures enclosed in glass.Instead use fabric pictures.

- Look at the noise reduction coefficient when choosing acoustic paneling. Sikorski says to choose ceiling tiles with a .9 NRC for an open plan (no hard walls) and .5 for a closed plan (with partitions).

- Do not choose wall panels that are too high. "Agents often want supervisors to know they are working, so they may talk louder so they can be heard," says. She recommends 42" panels, and never anything higher than 48."

Indoor Air Quality Considerations

The National Institute of Safety and Health sends investigators to evaluate potential health hazards in workplaces in response to employer, employee and agency requests. They refer to the term "Indoor Environmental Quality" to describe problems such as air quality, comfort, noise, lighting, and ergonomic stressers (poorly designed workstations and tasks). These are what NIOSH investigators typically look at to determine if there is an air quality problem.

1. Pollutant sources: Is there a source of contamination or discomfort indoors, outdoors or within the mechanical systems in the building?

2. The Heating, Ventilating and Air Conditioning (HVAC) system: Can the HVAC system control existing contaminants and ensure thermal comfort? Is it properly maintained and operated?

3. Pollutant pathways and driving forces: The HVAC system is the primary pathway. Are the pressure relationships maintained between areas of building so that the flow of air goes from cleaner areas to dirtier areas?

4. Occupants: Do the building occupants understand that their activities affect the air quality?

Repetitive Stress Injury

Well over 300,000 disorders associated with repeated trauma are reported to The US Department of Labor Statistics annually. They classify disorders associated with repeated trauma as conditions due to repeated motion, vibration or pressure, including Carpal Tunnel Syndrome.

According to Susan Fleming at the Occupational Safety and Health Administration (OSHA), an occupational illness is different from an injury because it occurs over time. "Repetitive motion injuries account for 60% of occupational illnesses," says Fleming. "They are the fastest growing in the workplace."

The Bureau of National Affairs, in a report entitled *Cumulative Trauma Disorders in the Workplace: Cost, Prevention and Progress,* reported that a case of carpal tunnel syndrome can cost as much as $30,000, and one third of workers' compensation is paid out for repetitive motion injuries.

When reps spend hour after hour at a workstation it's important to encourage exercise breaks and instruct agents on how to properly type.Encouraging preventive strategies and demonstrating proper techniques is the best way to minimize the risk of RSI.

"In 1970, OSHA established some usage guidelines to protect workers,but they never thought that the number of computer users would reach such a level in the nineties," says Gere Picasso.

"They are in the process of drafting new guidelines. Claims are now so substantial that keyboard manufacturers are placing warning labels on keyboards."

NIOSH researcher Dr. Vern Putz-Anderson, recommends using wrist rests but says they should only be used as a place to rest wrists in between sequences of typing. "When wrist rests are used as a place to hold arms while typing they can actually aggravate a condition," says Dr. Anderson.

"During typing fingers should float over the keyboard. Movement should come from the shoulders and arms because muscles in these areas are

HOW TO DESIGN AN ERGONOMIC WORK AREA

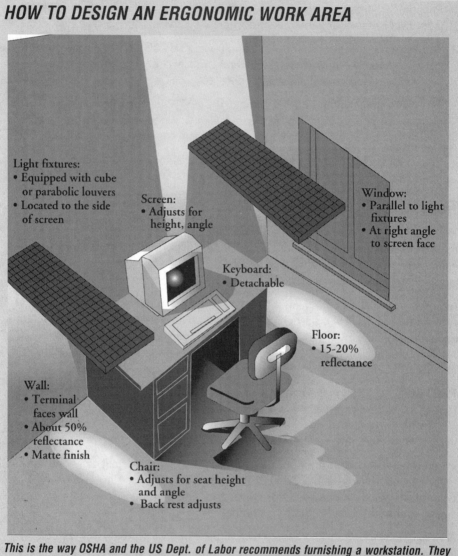

Light fixtures:
• Equipped with cube or parabolic louvers
• Located to the side of screen

Screen:
• Adjusts for height, angle

Window:
• Parallel to light fixtures
• At right angle to screen face

Keyboard:
• Detachable

Floor:
• 15-20% reflectance

Wall:
• Terminal faces wall
• About 50% reflectance
• Matte finish

Chair:
• Adjusts for seat height and angle
• Back rest adjusts

This is the way OSHA and the US Dept. of Labor recommends furnishing a workstation. They consider these suggestions general guidelines to minimize end-user fatigue.

more resistant to stain than the tender muscles in wrists." He says bending wrists also causes unnecessary stress since fingers are not elevated and forced to reach for keys.

Looking at other support centers

There's a lot you can learn by looking at what your competition is doing. Some of the most successful help desks have learned the value of sharing information. They realize that they can't afford not to learn what their competitors are doing to better serve customers.

Help desks know that when customers call, they are looking for answers to problems. Even if your technician realizes the problem does not have to do with your product, taking a hands-off approach and telling the customer the problem resides with another vendor's product does not suffice.

Trying to pinpoint the customer's problem (while not spending too much time or cost) will lead to a satisfied customer. That's why many help desks have joined forces. Instead of pointing the finger at another vendor it's a lot more effective in winning points with the customer to take an extra step, maybe ask a few more questions, make a phone call, or pass the problem on to the vendor responsible for providing answers. That will at least save your customer from making another phone call and repeating the problem.

Integration problems are the most common situations where help desks are most likely to work with another vendor to try and solve the customer's problem.

There's great value to be had in knowing how your competitors support their customers, and in turn, sharing your strategies. That's where benchmarking comes in.

Verity Consulting (Los Angeles, CA) puts together comprehensive benchmarking studies of corporate helps desks that includes Ryder,Novell, Chase Manhattan Bank and Teltech, just to name a few. Each company that participated in the study was required to answer questionnaires, complete telephone interviews and participate in around table discussion of results. Before the round table discussion, Verity analyzed and put together the findings of their research to form a "model" containing the key success factors for managing a world class help desk.

Help desks had the option of not disclosing their identity, but Verity's Jeffrey Bonforte says 75% declared themselves to other companies. "Help desks are saying 'can I afford not to learn if another company is learning?'" says Bonforte. "Competitive issues disappear after these companies meet and talk. Everyone goes to a higher level."

"Benchmarking opens people's eyes," says Bonforte. "It lets them see things from a different perspective. In a comprehensive benchmarking project, the statistical comparison is just the beginning. The real value comes from analyzing the underlying processes that drive the performance differences revealed by comparing metrics." Bonforte says benchmarking allows help desks to make changes at an evolutionary rate. He says the help desks that don't think they need to learn what other help desks are doing are usually the ones that are not very successful. "Unless help desks move ahead at the speed technology moves ahead, they will fall behind," says Bonforte.

Understanding the Support Process

Just because a help desk may answer all, or a large percentage of calls with a speedy response rate, does not make it a good help desk. "It depends on the quality of service and the desk's overall role in the IS infrastructure," says Bonforte. "The same goes for help desks that escalate only 10% or 20% of their calls. We studied one desk in Europe that escalated 40% of their calls, but provided top-notch support."

Bonforte says a help desk like WordPerfect can't afford to answer every call. He says it's common for half of their customers to not go through on the first call or two, but 80% eventually get through. He points out that this does not necessarily make them a bad help desk, but a normal abandonment rate should be more like 10%.

There has to be a balance between cost and quality," says Bonforte. "A three to ten percent abandonment rate is natural. Anything below that could be a sign that you are over staffed."

Bonforte says the help desks most likely to have the best automation are the ones who have been around for two or three years. Also, he says it's common for the smaller help desks to provide better support than the larger desks. "A bigger company needs more management infrastructure," says Bonforte. "A smaller company has fewer products to support or questions that tend to be simpler. The parameters of support are smaller."

COMPOSING A "BEST-IN-CLASS" HELP DESK

Here are seven of the many characteristics Verity Consulting discovered about the Best-in-Class (BIC) help desks:

1. The BIC don't try to be all things to all people. They rigorously define their role and set their customers' expectations accordingly.

2. The BIC manage themselves like an independent business. Help desks that ignore marketing and customer satisfaction (including senior management) are most likely to beoutsourced.

3. The BIC have integrated themselves into the IS value chain. The BIC are involved from planning to roll-out. Good support starts before a product or application is even released.

4. The BIC don't fight fires, they prevent them. They spend time off the line to improve their processes to eliminate problems at their root.

5. The BIC know if they're not measuring it, they're not managing it. They set a small number of balanced performance goals and track their performance aggressively.

6. At the BIC, management, not technology, drives practices and performance. They are successful despite technology, not because of it. Nothing can replace good management.

7. The BIC survey their customers regularly: The BIC let their customers' voice their needs and let those needs drive the evolution of the help desk.

Sometimes not enough value is placed on the help desk. It's important for the help desk to have an instrumental role during the development of products. Bonforte says the best companies have integrated their help desk throughout the whole process of building an application. "The help desk should have a hand in planning and developing applications," he says.

"The help desk should be heard through the whole process, not considered an afterthought. The customer will be frustrated if the help desk is just coming up to speed as they begin using the product. A help desk is designed for failure if IS believes support is simply something you tack on at the end of development."

While software prices may have dropped, costs for supporting customers have risen. Today free support is unaffordable for nearly every help desk and is almost unheard of. Bonforte says help desks can charge for support per workstation or per person but thinks it's much more effective to negotiate for support.

"Customers haven't learned that when they see a drop in software prices, support gets pulled to make the price lower," says Bonforte. "(Customers) want lower prices but complain when they can't get through for support."

"Customers should pay for support based on the percent of calls they generate," he says. "This way, if one department is placing a greater than average number of calls, the help desk has been tracking that and can inform them that they may be over using support. The desk and the department can then work together to figure out why, and how to lower their call volume."

Bonforte is not an advocate of free support. "If support is free customers won't be sure how much they can complain," he says. "When they are paying they can take a more active role. Companies should want to know how much time their employees are spending on the phone for support. Paying gives managers a way to track this. Putting a price on something is a way of informing a buyer and helping them to understand expectations."

"Customers should understand the support process," says Bonforte. He says Mercedes Benz, for example, might tell someone their car won't be ready for two weeks but every couple of days they will call to tell them what point they are up to in the job. "When customers understand the process they will trust you more."

He says you can automate this process by allowing customers to check your knowledge base themselves. Bonforte also thinks customers should be able to look at their own past problems to see how they were solved so they can try to solve the problem independently if it should arise again. This would save calls to the help desk, saving valuable rep time.

Outsourcing Support

If this all seems like to much to think about, you may want to consider outsourcing some support functions.

Support per workstation averages several thousand dollars a year. Faced with the daunting task of convincing upper management to invest money for improvements, many companies are considering outsourcing as an option to increasing staff.

Approximately 40% of help desks will have some form of outsourcer participation by 1998, according to Bill Keyworth of the Gartner Group.

Outsourcing some or all of the functions of your help desk can help eliminate headaches and save money if you don't go into it blindly. You need to develop a plan, choose the right service bureau and maintain an active involvement when it comes to how they handle your calls.

A service bureau will answer calls on behalf of your company, transparent to the caller that it's a service bureau answering the call. You can tell a service bureau how you would like them to handle calls. Tell them the type of software you want them to use, the service standards you expect and how often you want to be sent reports.

They'll send you information that tells you about the customer, his problem, how long it took to solve and any other information you ask for, at a price. I recommend receiving daily reports that identify the type of call, the length of each call, whether or not a callback was required (and if so, how long before the customer was called back) and the amount of time that passed before the call was answered.

Perhaps you've already invested in an automated call tracking system and problem resolution software so that you can handle more calls in less time. You're figuring it wouldn't make sense to outsource anymore.

But eventually your company will grow and more customers means more problems. It's a big expense to hire more support reps, buy additional user licenses, new workstations, etc., not to mention another salary on the payroll.

The roll-out of a new product is sure to generate an increase in calls on that

PROS & CONS

These are some potential benefits of outsourcing:
- ✔ Near-term cost reduction
- ✔ Long-term cost reduction
- ✔ Better IS expertise
- ✔ Improved business focus
- ✔ Improved flexibility

These are some potential risks of outsourcing:
- ✔ Loss of management control
- ✔ Cost escalation
- ✔ Reduced responsiveness
- ✔ Reduced flexibility
- ✔ Loss of in-house expertise

(Source: The Help Desk Institute.)

product. You may want to consider outsourcing calls in this area of your help desk rather spending time, effort and money training and hiring more reps to deal with the increase in call volume. What's nice is that you typically will pay a price per call or one price for a certain number of calls.

Say you're a software company that often releases upgraded versions of

Tips for Buying Problem Resolution Tools

Our experience helping customers automate their help desks has given us these insights:

1. Evaluate prospective help desk applications in as much of a real-world environment as possible. The proposed application should be installed, with real data, and evaluated by the support agents who will be using it. A realistic call volume should be processed through the prospective solution, using real data. In other words, put the application to productive use if at all possible before buying. Make sure it will deliver the kind of performance you need.

2. Pay close attention to the problem resolution capture and retrieval technology. Most help desk products do an adequate job of call logging and tracking - it should be easy to automate these functions. But a substantial part of the business case for implementing help desk automation is based on the ability to store and retrieve solutions out of a knowledge base. In actual experience, it can be hard to capture the problem solving expertise of the support agents in a knowledge base - you need this process to be as simple as possible. Your help desk tool should not erect any barriers to capturing this expertise. Make sure your support agents understand and are willing to use the tools in your help desk application to do this. If you are supporting proprietary technology for which no commercially available, prepackaged knowledge exists, it is imperative to carefully evaluate the knowledge capture and retrieval capabilities of any solution under consideration.

3. Pick a vendor who is willing to work with you to incorporate suggested changes and enhancements into their product. While there are many outstanding help desk products available today, few customers will find a product that does everything they want - a perfect match. Part of your decision criteria should be a willingness for the vendor to listen. Talk to their references to find out how responsive they have been to requests. Make sure that the product you buy has a history of evolving based on customer input.

4. Remember, the best help desk product is one you will implement and use. Don't put too much emphasis on which database it uses, what language its written in, or other technical specifications? Does it work? Do the customers who have bought it before you use it effectively? The most technically advanced help desk product available is of no value if you can't or won't use it. This sounds like oversimplified common sense, but I speak from experience here. We regularly sell our product to customers that have previously invested in other help desk software that never got anywhere with it. Above all else, place a premium on ease-of-use.

5. The product is not the process. Many buyers believe that the help desk tool they buy, once installed, will by default enforce the process disciplines necessary to run an effective help desk. This is not true. Tools can do much to help streamline a support process and make it more efficient and effective, but the tool in and of itself is not the process. Implementers of help desk tools should still define a process first, then find a tool to support that process. Failure to do this will result in disappointment with the outcome of the automation effort.

—Teubner & Associates

the product, enhancements and software add-ons. With each new release comes new users with questions. But call volume will often fluctuate leaving you faced with staffing headaches.

A service bureau is equipped to handle all the peaks and valleys. They use all kinds of call and workforce management software that few help desks currently have installed. That's just one less thing to worry about if you outsource these calls.

You'll also avoid the wrath of upper management when problems are not being solved fast enough or too much time is being spent on each call, driving up the cost of the help desk.

When you outsource you can offer live support 24 hours a day every day, or at least extend the hours of operation. Such accessibility can give you an edge over the competition in the customer's view, or allow you to charge a bit more for support.

Since all a service bureau does is handle customer calls, they are well-trained in this area. They do all the agent training and deal with the sticky issues like how to maintain high service standards when reps are out sick and are prepared for high turnover rates.

And they have all the latest technologies in place. For a small help desk, although technologies like IVR and computer telephony links help save money in the long run, there's still that upfront investment.

It's important to run a cost analysis comparing costs per call in-house versus what a service bureau charges. In the help desk that can get very complicated. Costs depend on whether or not staff members consult other staff members for calls. That's a hidden support cost that can be more expensive than you'd think — and it's hard to track.

You also need to look at service agreements, comparing what your customers pay each year to how much it costs you annually to handle all of

Tips for Buying Problem Resolution Tools

In buying a problem resolution tool, customer support managers should look for applications that allow them to take full advantage of the collective knowledge of their support organization. A support organization's analysts are their greatest resource and if the organization can capture and reuse those analysts' knowledge and immediately share it throughout the organization, they can dramatically reduce the time required to solve customer problems.

This is especially crucial for those organizations in the high-technology industry, where products change rapidly and there is continual pressure to solve new and increasingly complex problems. Buyers need to be wary of systems that do not allow them to capture and reuse knowledge in real-time, since such products will hamstring any increase in productivity a problem-resolution tool might otherwise provide.

— Primus

your help desk calls.

Then, if your help desk is not automated, you have to tack on the cost of solving the same problem repeatedly. Joe might find a solution, think that he should notify others, but forgets, so you have Jim, Jane, Peter, ... (you get the idea) solving the same problems time and again.

You can bet that doesn't happen at a profit-driven service bureau. They will be using the best equipment (they can afford to) and have effective procedures in place because it's in their best interest to handle the most calls in the least time.

Ensuring Satisfied Customers

Choosing the right bureau is of fundamental importance. Their agents will be representing your company and the impression they leave on customers will affect your business for the better or worse. Most outsourcers pride themselves on being the mirror image of the companies they represent. Here's a checklist to help with the selection process.

✔ Check their reputation. Talking to their customers can help, although they're sure to hook you up to their most satisfied ones. See if they'll accept visits unannounced. That way you can observe their operation when they're not expecting you.

✔ Hire them for a trial period to see how it goes. Towards the end of the trial (say two months) talk to the people who called them (or send out surveys) to gage customer satisfaction.

✔ Remember that you are paying an outsourcer as if they were part of your staff. You should feel like their employer. Clearly state your needs and make sure they understand your expectations. Remain closely involved. Check reports they send you every day.

IBM's Well Kept Secret

You might have known IBM handled support for products other than their

Smooth Sailing Tips

Most, if not all, problems, stem from basic management deficiencies. Most of these are in the people management area. The wrong type of person is hired - typically they are very technical but can't follow the rules and can't communicate with customers. Another typical problem is that the rep really wants to do some other job and thus doesn't take the support job seriously. This attitude is often reinforced by a company atmosphere that makes it clear that support is not important and/or customers are pests. Often CSRs are not given proper incentives, i.e., finishing "special projects" counts more than taking calls. One of the main benefits of a request tracking system is that it allows proper rep incentive plans to be put in place.

Occasionally the problem is a tool and technology problem. These problems are typically associated with home-grown help desk tools or old monolithic mainframe based products.

— Opis

own, but would you have guessed that 90% of their callers are users of their own products and services?

"Only 10% of our workload falls to IBM product support," says IBM's Bill Auvil. "The rest is end-user support for other companies." Auvil admits that it does not make IBM happy to hear him tout this statistic, but as the computer giant's marketing support and planning manager it makes him proud.

"For 99% of our clients, we are extensions of their VRU," says Auvil. "If someone hits 2 for PC shrinkwrap, for example, the call is routed to us, transparent to the caller." The VRU answers the call by welcoming the caller to the help desk and prompts them to enter their access code.

The toll-free number the caller dials in on (DNIS) informs IBM of the company the caller is from (for internal employee help desk calls) or the company the caller is trying to reach (such as a software company who they require support from).

For customers who prefer that callers do not go through an IVR system, they offer what Auvil terms "a live VRU" where "traffic cops" (which are operators) take calls and route them to the analysts best equipped to handle the caller's problem.

Auvil says they support 2,200 different hardware products from 300 manufacturers and they support over 700 general business applications. He is very clear when defining his priorities which have made them award winning support providers for the close to 1,000 mostly Fortune 2,000 companies who hire them.

Here are those priorities:

1 **Customer Satisfaction.** Auvil says they hire external surveyors ■ who callback ten percent of help desk callers every 24 hours. To not be repetitive, no one gets a call more than once every 42 days, unless they've had a problem.

"An unbiased person makes the calls and if someone seems neutral or dis-

Change Management, part of Impact from Allen System Group, lets users plan for, track and coordinate any changes.

satisfied they're told they can speak to a help desk manager," says Auvil. "A manger must return the customer's call within 15 minutes."

He adds, "Comments are not always negative. Sometimes customers want to speak to someone to offer a compliment." He says they don't send out surveys because the response rate is so low and by time someone fills it out they may not remember the kind of support they received.

Although IBM's clients may be pleased with the help desk service offered to callers, not all of their clients are motivated to come forward and attribute their help desk success to IBM. In a way it makes sense — Why should a Fortune 2,000 company admit that someone else can do a better job?

Even Auvil admits there can be negative connotations. He says that's why it's best that they are mainly an extension of the help desk —available to supplement existing support.

2. **To have an average hold time of two minutes or less.** ■ (They're currently averaging about 2.5 minutes.) Auvil says he was on the road 160 days last year listening to the goals of potential customers and other help desks.

Unlike some service bureaus, when a client wants to hire IBM as a help desk extension, IBM doesn't necessarily ask them the service levels they're looking to achieve. Instead they share their goals, which are high in the support services arena where it is not uncommon to call a help desk and remain on hold for 15 minutes or more.

Smooth Sailing Tips

The best advice I can give someone who has responsibility for a support operation is: define and track meaningful measurements. Help desk applications tend to collect a wealth of interesting and valuable data, but too few operations really exploit that data for the benefit of improving products, service, and customer satisfaction (an entire book could be devoted to this subject).

Our experience shows that virtually 100% of customers who buy help desk applications will implement call logging and tracking processes with their help desk software - its pretty easy. A smaller percentage, between 60-80% will also use their application to build a knowledge base - it's not as easy to do as logging calls. Finally, even fewer, I'd estimate less than half, get serious about using the data they capture to their benefit.

All the vendor products have some sort of reporting capability, and everyone who buys help desk automation has every intention of using it. Too few get this far in their implementation, however. Until measurements have been defined, are being tracked, and actions are being taken to improve them, you should not consider your implementation to be complete. Examples of these measurements include: responsiveness, product defects, workload distribution and others.

—Teubner & Associates

3. For any problem within their means they have a cycle time of **resolv-ing all open calls in 24 hours.** If it's a problem out of their reach, maybe residing with another vendor's hardware or software, they'll work with the vendor to keep the customer abreast, says Auvil. He says they realize that not everyone uses IBM hardware and applications so they are a multi-vendor shop.

4. **They strive to resolve all calls on the first call.** Sure, you're thinking that's what any sane help desk wants. But Auvil says they meet that goal 92% of the time. They average 1.3 calls per problem.

5. **Problem duration goals.** An average call resolution is less than 20 minutes. That includes an additional three minutes they spend ensuring information like the type of problem, time spent on the call, type of resolution and other information is provided for reports.

They're able to achieve such incredible statistics since each support rep supports only between three and five different applications.

Auvil estimates that a typical help desk spends $105,000 to $110,000 a year for each support person. He says IBM can typically provide better support for less than half of that amount. Skeptical companies often tell Auvil they don't spend that much, but when he points out over looked costs, his estimate turns out to be on the money or lower.

Auvil admits that outsourcing cannot be 100% remote. He has no intention of tying to eliminate the help desk manager. "There needs to be close ties and hand holding," he says. "Each of our customers has an IBM customer coordinator and we need a primary contact at the company so we can stay well informed. We want to be a true extension of the help desk."

Handling calls for so many clients while trying to maintain stringent service levels is no easy task. Auvil says you cannot outsource and forget to tell the outsourcer about major changes that will generate an increase in the volume of calls.

"One client made a major upgrade to their LAN and forgot to tell us," says Auvil. He says that morning practically every employee was phoning in.

But even a case like this won't necessarily turn into a disaster. Analysts have machines at home. If call volume really shoots up Auvil says these analysts can dial in and take calls.

"We like to have a phased roll-out when starting to handle calls for a new client," says Auvil. He says it makes for a smooth transition when they can start off handling 4,000 calls for the first month for a client who typically gets 30,000 calls. While they continue to seek high ticket clients like Ernest & Young, Miller Brewing and Chrysler, they are also targeting smaller companies.

They currently employ over 1,000 support reps of all different skill levels who work in offices in Atlanta, Tampa, Chicago and Dallas. If all reps are all on the phone at one office, calls can be overflowed to another office. Soon a new center in Charlotte will be up and running — employing another 350 support reps.

While the help desk at IBM may seem ideal, like any help desk, they too have faced problems, but fortunately remedied them before they had much impact. For example, they never used to evaluate the billing options clients chose.

One client told Auvil at the end of 1993 that he thought service was great, but by 1994 the company cancelled their contract with IBM. When Auvil called the customer to find out why, the customer said charges were much too high when they broke down what it was costing to solve each problem.

Auvil realized the customer was paying for unlimited problem solving and since they were not getting enough end-user calls the expense per problem was through the roof.

It made more sense to change the client's payment plan by having them pay one price for the handling of a small block of problems. Now IBM performs periodic financial reviews of each account's contract to ensure they have the most cost effective payment plan in place.

Auvil says they also recognize that burnout is high among help desk analysts. That's why they get analysts off the phone for a couple of hours each day and involved in other projects.

"They spend 6.1 hours on problem solving and the other 25% of their time is spent involved in training or special projects," says Auvil. "We try to make the help desk part of a professional career path."

Maybe that's why IBM continues to see success. Isn't the combination of a good reputation and effective strategies coupled with such personal touches an attractive draw to customers?

IBM recognizes their niche in the market — Auvil says they're not bothered

Manage-it from Baron Software lets users get customer information relating to a support calls whether using a standalone or network set-up.

by companies reselling their service. They may be selling service under their own company name but are using IBM's help desk service. Or maybe their key is lack of greed — they realize it's in their best interest to share the pie to see continued success.

The following is an example of another outsourcing organization that runs a support center that handles about one millions calls a year.

Softbank Services Group(formerly UCA&L). Based in Buffalo, NY, Softbank provides outsourcing services for telemarketing and support to companies like Intuit,Microsoft, Compton, Informix and Walt Disney.

Goals: To reduce support costs for clients and increase first call resolution. At the same time, they wanted to be able to use support reps who do not have very extensive technical training.

Significant Technologies: Astea's PowerHelp and Case1 for call tracking, problem resolution and case-based reasoning; FolioViews from Folio for full text retrieval; AT&T's Definity ACD and telephony services for computer telephony integration(CTI) so agents get screen pops; an IVR system from Voicetek that they've enhanced for debit card and prepaid support using Antarescards from Dialogic; speech recognition from Linkon; text-to-speech from Lernout & Hauspie; call center management software from Pipkins and in some cases TCS Management.

How the Application Works

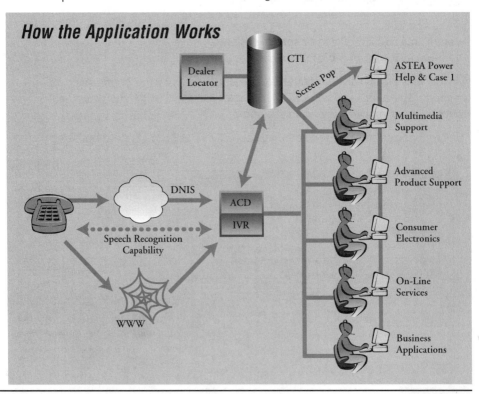

Results: Astea's Object Based Support Architecture was very important to Softbank because they needed to deliver customized support across multiple product lines.

"With about 1,000 campaigns and over 100 different clients who want different screens in different places, we needed OBSA to make it easy to add new systems," says Softbank's Paul Bandrowski.

Overall in 1995, Softbank handled over 10 million calls in their inbound telemarketing division. In their support division, they handled about one million calls. They have about 200 technical support reps.

Softbank has been able to achieve their goals: on average, they solve 80% of their calls on the first contact with an abandonment rate of about 4%. They answer 80% of all calls in under two minutes.

Calls are routed based on DNIS and selections callers make in their IVR system. When callers enter their selections (or speak them) into the VRU, the call gets sent to the appropriate agent group along with a screen pop. PowerHelp is installed at some client sites so information can be sent to the customer site.

They're working on a setup that would allow Web crawlers on the Internet to be transferred (via a modem line) to a live agent and a dealer locator application for the Internet. Currently callers can find the dealer closest to them — using text-to-speech the system reads them back the names.

Tips for Buying Problem Resolution Tools.

Since Help Desk software is being applied in a myriad of ways, look for a vendor that knows your business.

Don't let the computers and software take the decision making capabilities away from the customer service reps.

Don't get a system that is so complicated and hard to use that the customer service reps focus on the software instead of the customer.

Get a system that allows the customer service rep to function even when the network or workstation is down or a customer is just very upset. This does happen even in today's advanced technical environment.

This means that:

A manual tracking process should dovetail with the software to allow for entering the information at a later time if the computer system is down.

The customer service rep can use the time to defuse the customer and take notes while the customer is talking then enter the information after the call.

Get a system that allows you to spot chronic problems or chronic complainers. Both of these cost you money.

—Oasis Technology

➲ Chapter 3: Deciding Whats Right For You

Choosing the Right Help Desk System

When it comes time to choose a help desk system, it's easy to get lost amid the throng of products and features. It's probably the most serious decision your help desk will make. When you are spending thousands of dollars (or hundreds of thousands) to automate your help desk, the last thing you want to do is rush a selection. You don't want to find out later that you've picked a product that lacks some instrumental features. With so many product choices from so many vendors making a selection is tough.

Take the feature so many vendors now tout: "easily customizable." Two vendors, both making this claim, may have very different ideas of what it means. Typically, the less you spend the fewer features you'll be able to customize to look the way you want.

"Years ago you could not touch the screens," says consultant Albert Starck, president of Hyperion Associates (Fredericksburg, VA). "How it came out of the box was how it had to look." Today most systems are customizable to varying degrees. Confusion over customizability is typical of the buying process.

Here are some of the steps you should take to make the selection process as painless as possible.

1. Figure out what departments can take advantage of the software — right away, and farther down the road.

"A help desk system can benefit the sales, payroll and human resources departments," says Starck. "Four years ago a product was bought as a tool for only the help desk. Now the automation is being used enterprise-wide."

Help desks often keep tech support and order entry information separate and end up with multiple customer databases. The support rep should be able to log a problem and without leaving the system, place an order for a spare part, for example. There needs to be a single point of access.

2. Compare your requirements with what the product offers. Choosing has become harder because of the huge shift of vendors going from DOS to Windows and from file server-based systems to client/server-based systems.

Starck says that a couple of years ago choosing between DOS or Windows and file server or client/server were big considerations that went into the selection process. That's no longer the case.

"The biggest cut point in the selection process comes down to price," says Starck. He says help desks look at the cost of the software, costs for training, maintenance and installation when comparing product prices.

Product Scoring Matrix

Problem management						
Ease-of-use						
Problem description: summary, detail activity						
Problem entry via e-mail (auto, problem log)						
Clear problem history						
Problem classification levels						
Query and diagnostic features						
Common/prior problem lookup						
Key word searches						
Symptom related lookups						
Assignments, referrals, & escalation						
Problem priority levels						
Automatic escalation via e-mail notification						

Assign values for exceeds, meets, work-arounds, fails to meet. For example:
9 - Excellent implementation with added value
3 - Satisfies feature
1 - Borderline implementation (or there's a work-around)
0 - Feature is missing
Here are some examples of features you might want to assign values to as part of your selection process. You can use this chart as a guide, substituting, deleting or adding features important to your help desk. (Source: Hyperion Associates, 703-899-2985).

You'll probably narrow down your choices to five out of 20 or 30 products you may have considered. The hardest part is deciding which out the five products you seriously evaluate is the right one for your help desk.

Another factor that goes into the selection process is the database the product uses (such as Sybase or Oracle). Also, integration with knowledge tools can be a strong point if you don't want to build your own knowledge base from scratch.

Put together a project team. Starck suggests composing a team of no more than 10 to 15 people who will rate different products. Put together a list of must have features, nice-to-have features and okay features.

Starck says it's important to get input from technicians and experienced users. "You need to have someone familiar with using an automation tool and know what to look for," he says.

4. Consider ease of use. "A product can be easy to use but difficult to learn," says Starck. "Certain products are known to be difficult to implement."

What you can do with a product right out of the box is a good indicator for how much you will need to put into the product before you are able to use it.

If you need to add code to pull up client histories, for example, you probably will need to spend time writing code for other things too before you can be up and running. Starck points out that it's especially important to choose a product that's easy to use since turnover is so high among help desk employees.

"Some features like multimedia may be overkill," says Starck. Although he admits such features demo nice, he questions their practicality. He advises choosing based on requirements. "There may be bells and whistles you don't need and make the product hard to use."

5. If you still can't decide.... test two or three products after eliminating those that don't make the final stages, and rating these systems on each feature and function during the software evaluation.

"When testing, if an important feature is not there, call the vendor," he says. What's lacking could be something simple they'll add for you, or the capability might already be there but the vendor must show you what to do.

"If you are down to a couple of products and one doesn't stand out above the other, consider documentation, vendor support and training," says Starck. "Call the vendor to see if they have a customer in a similar situation using their product." He even suggests site visits.

If you are reading this after you purchased a product that does not meet your needs, don't shelve it just yet. Pinpoint what it is you don't like and speak to the vendor.

"Make sure you have the latest version," says Starck. "One company I dealt with was not happy with the product they were using, but they were seven or eight releases behind."

On the other hand, if you are using a system in-house developers built from scratch, and are disappointed, you are not alone. Many help desks try developing and using home grown systems, perhaps in an attempt to get away from an archaic paper-based system. But it's hard to develop a system that does much more than basic call tracking. "It's expensive to build your own system," says Starck. "And to put in features like auto escalation requires a lot of programming."

Best advice? Scrap your home grown system and buy something from a vendor who specializes in help desk systems and whose product contains all the latest features and functions. You'll also have the satisfaction of being able to yell at someone other than your own employee when things go haywire.

10 Steps to a smarter purchase

Here are some steps to follow to help you make the best purchasing decision you possibly can:

1. Form a project team.
2. Define project parameters.
3. Develop a requirements list.
4. Organize and prioritize requirements.
5. Conduct market analysis.
6. Conduct comparative analysis.
7. Score candidate products.
8. Assign performance scores and analyze results.
9. Select the best product.
10. Present findings to project team.

—*Source: Al Stark, Hyperion Associates.*

Creating an RFP

The time has never been better for you to get help desk software with all of the features you want. In such a competitive market, vendors listen closely to what help desks ask for, doing their best to fulfill your every need in an effort to please and stand above their competitors.

Once you've figured out what features and functions are most important to you, what you can and can't live without and the approximate price you're willing to pay, you need to start sending out your request for proposal to the vendors who made the first cut.

The following checklist comes from Tammy Kirk of the Gartner Group and entails information you should consider when putting together your RFP.

Introduction

Background on the corporation and/or business division

Background on the project including committees or teams

Statement on the confidentiality of the information

Project Detail (reasons and motivations for the project)

Statement of business problem with a business perspective

Statement of mission/vision

Statement of scope and objectives

Project Schedule

Vendor RFP question deadline

Vendor analysis meeting (optional)

Proposal due date

Vendor demonstration day (optional)

Final decision

Proposed implementation start

Business Requirements

Detailed business requirements

Firm technical requirements

Installation, training, start-up and maintenance requirements

Project management and reporting requirements

Performance criteria for success of the project and solution

Some indication of performance penalties

Technical Environment

Current technical environment (e.g., hardware and software)

Technical architecture, including potential future changes

Proposal Requirements

Contents of the proposal, including references and key
    vendor personnel

Proposal format and number of copies

Guidelines on how to present costs

Documents to submit to establish vendor financial stability

Selection Criteria

Determining Needs — One Manger's Experience

The following comes from excepts of an article written by James Litchfield a help desk manager at a computer billing software company. Litchfield wrote this article with the intention that "neophyte help desk mangers might benefit from my experience" in building a help desk.

VISION. Once you've got most of your questions answered, it's time to discover the vision you have for your help desk. Determining your vision will have a significant impact on the strategy you develop to implement your help desk solution. This is the where you interpret your company's vision statement. When you think you have a pretty good handle on the company's vision, talk to senior management. Are you in tune with current company philosophy? If you are then you are well on your way to developing your vision of customer support.

On one end of the vision spectrum, you may decide that your vision is to focus on re-creating an old and failing help desk strategy. Perhaps you remember the seventies, when companies developed an Information Systems (IS) Department as a tool to support an efficient, albeit limited, environment of "dumb" terminals. For help desks that try to hang on to or re-create this environment, there is grave danger. These organizations will ultimately fail to provide adequate support to corporations with increasingly complex technical environments. At the opposite end of the vision spectrum, you may decide to focus on re-engineering help desk strategy. Within this vision is a customer support strategy that links new technology, high tech staffers and smaller financial requirements. You are not only concerned with the "most bang for the buck,' but also, "Will this technology hold up in the long term (5+ years out)?" In between these extreme points on the vision spectrum lie a myriad of strategies. Which one is the best? Only you can answer that question for your company. At my present com-

Smooth Sailing Tips

"It is very important that the problem resolution tool be appropriate for the user. The way in which the user interacts with the system must be appropriate to the level of the user's prior knowledge. For example, in a Tier 3 support organization, users need a tool that supports sharing experiences among equally knowledgeable people. Therefore, they need to be able to input and retrieve unstructured, anecdotal information about new and usually previously unresolved problems.

In a Tier 1 (1st level) support organization and for the end-user/customer, that model is not appropriate. These users need a very structured knowledge base that can automatically determine the appropriate troubleshooting strategy and can guide the user directly to unambiguous advice. This is only appropriate if the problem has been solved in the past. Understanding that not one tool fits all needs and fitting the problem resolution tool to the user is of primary importance."

— Eileen Weinstein, VP Operations, Advantage kbs

pany, Senior Management opted for re-engineering help desk strategy to the point of diminishing returns. That is, we wanted new technology at a reasonable price, and a technically qualified staff, based on an analysis of sales and projected sales. We wanted a help desk that will grow with the company into the 21st century.

SOFTWARE RESEARCH. Now that you've solidified your vision, it is time to discover what's driving your train. Is it hardware? Software? Personality? After several long discussions with my Director of Operations, we determined that software should drive our train. We decided that our strategy should be:

1. Buy the appropriate software

2. Buy hardware to support the software

3. Find people and train them to use the current configuration of hardware and software

What is "appropriate software?" Only research will answer this question.

What are you researching? You are researching how well any given number of help desk software packages satisfy a set of Functional Specification Requirements (FSRs).

What are FSRs? FSRs are simply a set of criteria you establish that are required to enable your vision to come to fruition. You might have several sets of criteria as well as several levels. Here's how I did it. The first step was to make a list of fourteen initial requirements that I thought help desk software ought to satisfy. I didn't just pull this list out of the air. It is based on the questions I asked, my experience, technical advice and advice of senior management and customers. This list contained such items as:

✔ The type of server I wanted to port the software to

✔ The type of coded fields I wanted to identify

✔ The type of reports I wanted to generate

✔ Customer information tracking

✔ Demonstration version of the software

Smooth Sailing Tips

Support centers can run more smoothly if there is a clear prioritization of tasks. Not all user requests can be met within an hour. Also, it is useful to distribute FAQs (answers to frequently asked questions) to users. Centers can run more smoothly if they automate and set up the best applications in the first place. Proper employee training is essential.

— Avalan Technology

After I listed these initial requirements, I divided them up according to "must have's", "should have's" and "nice to have's," giving a weight to each category. After listing these requirements, I made a list of software products based on a random survey of help desks within my company's industry. Then I set up a matrix of software products and requirements. As I matched requirements to software, I weighed how well the software met the requirement. This meant that I had to see, or conduct, a demonstration of each of these software products. This is the point in which you may want to reconsider the number of software products you analyze. I chose forty products and gave each product four hours of research time, keeping in mind that I am not trying to gain a working knowledge of each product. To be fair to each product, all "no" answers were validated with that product's technical support.

At the conclusion of this research only four products remained. I had rejected all software products that did not meet all of the initial functional specification requirements. At the conclusion of this step, if you have not eliminated at least ninety per cent of the products, you might want to consider more stringent requirements. Throughout this research, keep in mind that no value judgments are being made about the software.

At this point there is no "good" or "bad" software nor am I concerned with price. I am only concerned with whether or not the software met the requirements and how well the software met the requirements. The second step was to make a more stringent list of requirements that I thought help desk software ought to satisfy. This list is based on the requirements necessary to accomplish my vision for my company's help desk. This list contained such items as:

✔ Are the screens built with a 4GL language?

✔ Are the search engines built with artificial intelligence?

✔ Does the product have a custom report writer?

✔ Is the product supported by a seasoned support staff?

✔ Does the product utilize "pull down" menus?

Tips for Buying Problem Resolution Tools

Savvy help desk managers considering a help desk automation system should be sure to examine the full life cycle cost of each technology. It's not only the ticket price that counts, but also the effort required to maintain and update the knowledge base as you use it. Expert and case-based reasoning systems are acceptable in environments where the questions coming to the help desk seldom change. But in most situations, where questions change at warp speed, those technologies require dedicated programmers or case builders to try to keep up. This adds significantly to your cost. Managers should look for systems with built-in mechanisms to capture and maintain knowledge as a normal process of help desk staffers doing their everyday jobs.

— PLATINUM Technology,

Each response was weighted with a "**1**" for inadequate; a "**2**" for adequate but weak; and, a "**3**" for adequate. At the conclusion of this research, I had a numerical ranking of the remaining four products. This numerical ranking only tells me which product best satisfies the requirements I listed.

REFERENCES. The third step was to ask each of the remaining software companies for two references. I found most software companies are eager to provide references. Make sure that these references are in the same industry as your company (or at least in a related industry). I developed a two page questionnaire for each reference using the same questions. The purpose of these questionnaires was to determine if the reference had any issues with installation, documentation, ease of use, customizing, technical support, training and staff interfacing. You would be surprised at what people will tell you. I found it best to call each reference and set up an interview time.   As a side note, when I called each company, I also asked for sample job descriptions for support staff and supervisor positions. Most support managers had no problem in sharing these with me.

STRENGTHS AND WEAKNESSES. The fourth step was to compare strengths and weaknesses of each product based on the results of the questionnaires and the functional specification requirements matrix. My initial draft of this step was four pieces of scrap paper with the product name on top and a line drawn down the middle, strengths were placed on the left and weaknesses on the right. This step is pretty straightforward and does require some judgment calls.

REQUEST FOR PROPOSALS (RFP). The fifth step was to write Requests For Proposals (RFPs) for the remaining software and then evalu-

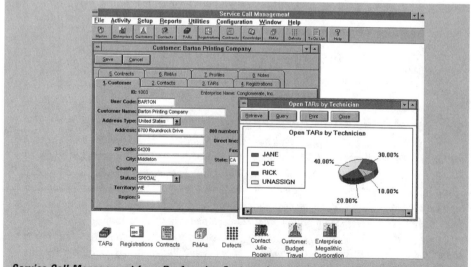

Service Call Management from ProAmerica Systems has a knowledge base, Query Lists, and a "To do list tool bar." The SCM Toolkit lets you customize existing screens.

ate the resulting proposals. It is crucial during this step that you make sure you are comparing apples to apples. Each proposal must be as identical as possible to the others. Vendors will list some items as "no charge" or "free" while other vendors will charge for an identical item. Regardless of the cost (or lack of cost), make sure all lists of items are as identical as possible. On two occasions, I had to request additional proposals with specific requirements that certain items be included on the proposal.

FINAL REPORT AND RECOMMENDATION. The sixth, and final, step is to write a final report and recommendation. I found that this was the easiest step because I already had all the statistics, charts and other data. I just needed to fill in the blanks with words. My report was four pages long with eight appendices containing all my evaluation tools. The first paragraph stated the purpose of the report. The next five paragraphs stated the results of each step of my research. This was followed by a discussion of each of the final products, a numerical rating based on the final requirements evaluation, a price comparison based on the proposals, a description of the trends in industry standards for support software, a conclusion and a one-sentence recommendation. Remember to keep your recommendation to just one sentence. You don't need to justify your recommendation, you should have already done that in the body of your report.

HARDWARE Once I decided on the software, I found that the software dictated the hardware requirements. We chose a client server-based product that had the following hardware requirements: - Standard IBM-compatible, pentium-based motherboard - 24MB RAM - 256KB Cache - 1GB Hard Drive - 14" Monitor - 64 bit PCI, 2MB DRAM, video card - 4X Multi-session CD Drive - Sound Blaster 32 IDE stereo card - 28.8 Fax modem.

PERSONNEL Earlier, in this article, I talked about calling vendor-supplied references to complete questionnaires. I specifically requested references within my industry so that I could borrow job descriptions and customize them for my company.

HINTS Do not prejudge any software. What was number one on my list may not have made the final cut for another company.

Save price comparison for last. You'll feel better knowing that a $10,000 product does the same thing for your company as a $50,000 product and at a fifth of the cost.

Don't succumb to high pressure sales tactics.

Ask for more hardware than you think you'll use in the near future. Remember the axiom, "Software grows exponentially to fill the space provided."

Hire technically competent "people-oriented" personnel. Another axiom to remember, "Help desk employees must have: the patience of Job, the wisdom of Solomon, the manners of Emily Post and a healthy horse sense."

CONCLUSION   In re-engineering the help desk strategy, support managers should consider these helpful insights: Re-engineering the help desk strategy is not a project but rather a living works, always evolving and adapting to the ever-changing technical environment.

Everything cannot be accomplished today. Save your sanity and focus on doing a few things well before you go home tonight.

Your customers will not use your help desk if they do not believe in you. Practice instilling confidence in your customers.

Campaign to win the hearts and minds of your company's personnel with sustained successful support efforts.   This will heighten company respect for your help desk and ultimately keep the vision alive.

Tips for Buying Problem Resolution Tools

Look for tools that provide as much out-of-box support as possible and ease of customization.

You have problems. The customer needs improved support service. You are choosing tools that can help. If the solution takes a long time to implement, what does that cost? The cost is not just the direct expense involved with the implementation, but the opportunity cost from not improving things faster.

Look for tools that provide high integration with network and systems management tools.

By uniting help desk applications with networking tools, your support function is better able to respond to end users' needs. The help desk can get early warning of problems before users begin calling the help desk. Once the help desk begins working on problems, whether from a network alert or from a phone call, important information about the user's PC and networking environment is available to aid problem diagnosis and resolution.

Look for tools with expert system technology supporting for first level and end-users.

Most organizations want more and more problems to be managed at the first level. Expert system technology along with network intelligence may allow 1st level to respond to 70 - 80% of problems. A key attribute of the expert system tool is that it should learn automatically and NOT require a lot of maintenance to engineer the knowledge (such as building cases in a case base tool). After all, in a changing environment you don't want to wait for the first level to be able to handle those problems.

The direction for many organizations is end-user self help. An easy to use expert system search tool is an obvious part of providing that self-help. This means end-users type their query in natural language and are not required to build QBE lists or restricted to specific key words they may not know. An important feature is to be able to partition the knowledge the end-user can obtain. This recognizes that there are many problems you do not want end-users to try to fix or are unable to fix.

Look for tools that provide a high degree of automation.

One specific automation feature that is generally not discussed is for the help desk tool to mine for trend problem data out of the database and present that dynamically to help desk users. The help desk can see when problems are developing.

— Peregrine Systems

Real Life RFPs

I picked out two RFPs submitted by help desk vendors for this book because I found them to be very comprehensive and good examples of guidelines you may want to follow when writing your own. The requirements these organizations had may be different from some of your own require-ments but can still help give you an idea of the questions you may want to ask for yourself.

The following RFP consists of requirements and questions posed to help desk vendors by a government agency looking for a system for their inter-nal help desk.

General System Requirements

System Must:

✔ Provide integrated help desk, problem, change and inventory/asset man-agement functions.

✔ Database should be non-proprietary, both on and off the mainframe and should include, but not be limited to the following: DB2, DB2/2, Oracle, DB2/6000, XDB.

✔ Provide for distributed and/or centralized data.

✔ Provide facilities to easily customize screens, panels and fields.

✔ Provide online help.

✔ Provide and maintain a complete audit trail.

✔ Provide automatic fill-in capability.

✔ Be able to automatically open and escalate problems.

✔ Provide 3270 access, as well as Windows and Presentation Manager (if a combination of presentations are used, should still be able to access any data as if using only one.)

✔ Provide security within the system.

✔ Provide varied authority levels.

✔ Provide the ability to search on any field and any concatenation of any fields.

✔ Provide automated alert/notification.

✔ Provide trend analysis.

✔ Provide cross platform functionality.

✔ Provide consistent applications across platforms.

✔ Not use a proprietary report writer.

✔ The data must be recoverable to the point of failure.

✔ Provide ability to archive data.

✔ Operate in a multi-vendor, multi-platform environment.

Service Desk Requirements

System must provide:

✔ The ability to quickly open service desk tickets.

✔ Access to system through 3270 and GUI presentation.

✔ The ability to automate fields.

✔ The ability to store service desk data on the LAN and/or mainframe.

✔ The ability to escalate service desk calls to problems without having to re-key data.

✔ The ability to open change records from the service desk panel without having to re-key data.

✔ The ability to easily customize panels and fields and have that customization protected in future releases of the product.

✔ Expert knowledge base function to provide quick problem resolution.

✔ Robust search capabilities such as "query by example".

✔ Automatic date and time stamp of each entry in the system.

Problem Management Requirements

✔ The ability to track problems by all fields or key words.

✔ The ability to validate and make certain fields required fields.

✔ Cross reference between problem number, change number and service desk call number.

✔ The ability to interface with automated scheduling, job management and performance management systems to automatically open and escalate a problem.

✔ The ability to perform trend analysis on problem causes and solutions history.

✔ The ability to easily customize screens and fields.

✔ The ability to perform automated escalation and notification based on multiple criteria.

✔ The ability to access configuration data from the problem panel.

✔ The ability to open a change record from the problem panel.

✔ The ability to validate input fields.

✔ Free form text.

✔ The ability to search the free form text.

✔ Real time updates, regardless of whether the problem was created on the mainframe or LAN.

✔ The ability to support a large number of records.

✔ Ad-hoc reporting capabilities.

✔ Database date and time stamp of each entry ticket.

✔ Ability to automatically track problems that occur from changes.

Change Management Requirements

✔ Online change submission, approval and status updates.

✔ The ability to track all changes with a given component or class of components.

✔ The ability to track relationships between changes and activities.

✔ The ability to copy an entire change record and all its related activities.

✔ Audit trail and history of all changes.

✔ Multiple authorization or approval fields.

✔ The ability to create/open a problem record from the change record.

✔ Service request capability.

✔ Robust search capabilities of free form text.

✔ Work flow management

✔ Inventory/Asset Management Requirements:

System must provide:

✔ The ability to track equipment through its life cycle.

✔ Configuration of hardware and software components.

✔ The physical and logical parent in the configuration.

✔ The ability to track all hardware and software components regardless of location.

✔ Service details for inventory items.

✔ Financial details for all inventory items.

✔ A central repository for entire inventory, including mainframe, PCs and LANs.

Chapter 3   **57**

✔ Ability to do trend analysis reporting on problems related to specific inventory items.

✔ Invoice tracking.

✔ Validation of input fields within the module.

✔ Free form text search capability.

✔ Entry and tracking of mainframe, PC and network software.

✔ The ability to automatically update inventory records from auto discovery tools.

✔ Real time updates for inventory items.

✔ Ability to run search arguments on all hardware and software components.

✔ Ability to tie problems to inventory components and vendors associated with those components to provide vendor management capabilities.

Automated Messaging

System must provide:

✔ The ability to interface with Automate 4.0, Omegamon, Netview, and IBM LAN Network Manager.

✔ The ability to automatically notify personnel of problems via fax, phone or beeper message.

✔ The ability to determine notification method based on pre-defined individual preference.

Reporting Requirements

✔ Provide standard reports with base product.

✔ Provide flexibility to tailor reports to specific needs.

✔ Provide ability to report on any or all data elements.

✔ Provide ability to move data into graphical presentation format (i.e. Lotus, Excel)

✔ Provide reporting capability from both the mainframe (3270) and LAN (GUI) presentation.

✔ Provide capability to generate time sensitive reports automatically.

✔ Provide real time update of reports.

Product Background

✔ How long has product been in existence?

✔ How many upgrades/enhancements have been offered since the initial release?

✔ How often are product upgrades provided?

✔ Is there an additional charge for product upgrades?

✔ What language is the product written in?

✔ Provide a list of customer references.

✔ Are there additional charges for user manuals and documentation?

Support/Training

✔ What types of support do you offer for your product?

✔ How many people support your product?

✔ Do you offer training on your product?

✔ What training is recommended for a casual system user.

✔ Do you offer training classes on a regular basis?

(This RFP was submitted for use in this book by Allen Systems Group, Naples, FL)

Smooth Sailing Tips

Streamline and optimize you current manual process before you automate. Automating a mess will only result in an automated mess. This situation will in turn result in a confused software vendor, a frustrated implementation team, cost overruns, and unhappy management.

Formally write up the problem that you are trying to solve and what your benefits will be. Be sure to include your hard dollar benefits as well as your non-tangible "efficiency" benefits.

Develop and refine a paper flow of how you want your new system to work. Keep your needs simple at first then add your wish list later. Sometimes putting too many requirements onto a new system at implementation time just results in a system that is more complex than it should be to fit your needs.

Submit your needs to various vendors and select two or three that you want to talk to. Don't make your evaluation just based on the demo diskette. Take time to talk to them and develop a relationship. The vendor could have some advice that will may help you.

Use your selected vendor to help "tune up" your plan.

Keep customization to a minimum at the beginning.

Implement in small concrete steps.

Keep your user community appraised each step of the way. This is the area that most implementers fail. As a result, the user community may feel that the new system will eliminate their positions or if they are not computer literate, they may feel threatened by the new system. In any case, their passive "non-acceptance" may spell doom for your new system.

— Oasis Technology

The following RFP comes from a network of hospitals looking for a help desk system. It is not the complete proposal, but an edited version targeting the company's main requirements.

Introduction

We are researching to find a problem tracking/help desk system that can be accessed through the enterprise-wide network. The software needs to be easy to use and cover the needs of all our users. The software must also fit within our existing environment of systems, including mainframe-3090 (MVS), Tandem, R56000 (AIX/UNIX), Novell LAN (Ethernet) consisting of over 3,000 devices on 40 file servers, or AS/400. This software should be a Windows-based product.

We have approximately 50 users on the current help desk software, but this number could grow to 2000 (non-concurrent) users. Our current average call volume is 97 calls per day, which has increased from 57 calls per day in January of 1995. This volume is expected to continue to increase. Multiple support levels utilize the current problem tracking software to track and resolve problems, supporting 5,000 employees

Nine companies were selected to receive this RFP.

Contract Cancellation

The Hospital Corporation reserves the right to cancel any or all contracts resulting from this RFP within 30-days written notice, with notification provided to the vendor by registered mail, at no cost to us.

Cost Liability

Central Iowa Hospital Corporation assumes no responsibility or liability for costs incurred in the preparation or submission of any proposal by any vendor, or for any tours, demonstrations of software, or consultations provided by any vendor.

Evidence of Ability to Perform:

Each vendor must show evidence of their ability to perform the responsibilities laid out in this RFP, through completion of Attachment A and attachment of other relevant information about the proposing firm.

Finalists must be willing to visit the Medical Center campus for an on-site demo of their product, and be willing to provide a 30 day evaluation copy of said software for testing purposes.

The vendor must describe, in writing, the types of support offered before, during and after implementation of said software, along with the cost.

The vendor must demonstrate their viability as a company by providing a narrative overview of the firm, including:

1. History of the vendor or local firm, including longevity of operation and affiliation with any branch or home offices. If branch offices will provide back-up for service, or technical assistance, this must be clearly stated.

2. Qualifications of the firm, and qualifications of sales and service personnel proposed and/or assigned to the hospital account. This includes level of education and training programs provided for service personnel.

3. Description of the service department, including number of service people, radio or phone dispatch, service ratings and awards received for the Service Department, etc.

4. A copy of the firm's annual financial statement for the most recently completed and audited fiscal year, attesting to the financial! stability of the firm and/or branch office. (If local firm represents a national vendor, we are interested in the local firm's financial statement, as well as the national vendor.)

5. Other pertinent information deemed appropriate by the vendor.

The vendor must demonstrate their ability to perform by providing the names, addresses, telephone numbers, and contact persons for at least three other businesses, organizations, or medical facilities of a similar size and with similar volume requirements at Hospital Corporation. References must be for current accounts, with software placed during the past twenty-four months.

Smooth Sailing Tips

Enterprise-Wide Access

The help desk is transitioning into the corporate service center, handling requests from reporting traditional IT problems to request about HR policies. To make the transition to the corporate service center requires deploying a fully open system. It means providing interfaces to a broad set of external problem entry and problem solving tools (e.g., e-mail, Web or Lotus Notes access).

Forward deploy means providing the right way for end-users to interact with the help desk. For some individuals or organization, the right way is e-mail, for others it may be Lotus Notes, still others may prefer to use a Web client. The choice will be based on preference and what function they are performing. For example, e-mail has obvious limitations for problem resolution. Again, the context is not just being able to get your problem into the system. It is a way to get information, search for information using expert knowledge tools, (since you may not know exactly what you are looking for), and help yourself.

Advanced notice from the help desk to end-users about problem status cuts down on the storm of calls about common problems. It also can make the help desk look more proactive to its customer community. A little self promotion can go a long way.

— Peregrine Systems

References must be provided for both software placement and maintenance service. (Three references are sufficient if the vendor placed the software and provides maintenance.)

Specifications for System Functions

The following is a list of functions that we desire in the system in the order of importance to us.

1. The system should have problem description and resolution fields which can be continued to contain unlimited text. Data entry functions for the system should include word wrapping, text editing, and spell checking.

2. The system should have customizable screens, fields and function keys. It should have user definable help functions and decision trees.

3. The system should assign an incident number to each problem at the initial entry of data to the input screen.

4. The system should allow us to enter a schedule of on-call personnel in a calendar format for first and second level technical support. Paging should be performed automatically by the system with reissuing of pages occurring after a user definable period of time when no change to the problem record has occurred. Also the rules for problem escalation should be user definable.

5. The system should allow us to query and report on all data fields. In addition, reports should exist for the following:

   ✔ all unresolved or open problems

   ✔ all problems assigned to a specific person

   ✔ all problems opened by a specific user

   ✔ all problems assigned to a responsibility area

   ✔ all problems for a specific device or device type.

The system should allow us to save all user developed reports and queries. The choice of displaying or printing the information should also be given.

6. The system should log each update to a call with a user ID, date, time and the change.

7. The system should allow for easy backup of data and the ability to archive old problem records to a history file. The history data should be available for reporting.

8. The system should include a security function that monitors user login and allows control of who can update data on a field level basis.

9. The system should allow for mass updates to the database so that all

occurrences of a specific data element can be easily changed.

10. The system should allow for definition of an interface to an inventory system that would maintain device inventory and map that information back into the problem record.

11. The system should flag all escalated calls and notify responsible people of the problem.

12. The system should allow data to be exported or imported from some of the popular database structures such as Access or dBASE.

13. The system should interface to a telephone database so that callers can be identified by telephone number and demographic data can be mapped into the problem record.

Format and Supporting Documentation

Each vendor is to submit their proposal and supporting documentation in the following format:

Response to this RFP shall be submitted in a notebook, clearly identifying the vendor or firm, tabbed appropriately for each section:

Section 1. Vendor Information

Attachment A - Vendor information Supporting narrative description of vendor

Section 2. Contracts

Sample copy of Purchase Contract Sample copy of Maintenance Contract Copy (or sample copy) of any sub-contracts between the vendor and a secondary (supporting) vendor Software License Agreements

Section 3. Billing

Narrative overview of billing procedures and/or billing department of vendor Sample of Purchase Invoice Supporting narrative description of purchase invoice Sample of "Exception" invoice Supporting narrative description of exception type bill

Section 4.

Please respond to the section on 'Specifications for System Functions' by number.

Section 5.

Other pertinent information deemed necessary to assist the Hospital Corporation in reaching a decision regarding problem tracking/help desk software.

(This RFP was submitted for use in this book by DataWatch Corp.)

Building a Knowledge Base

As a help desk with trained technicians you probably have the resources to find answers to customer questions. But you need to ask yourself where these answers lie. Chances are if your not using automation, you will look to experienced technicians, manuals, binders with written information, perhaps an online bulletin board and maybe even post it notes.

Using any one or all of these methods may eventually lead you to a solution. But how long will it take? And will someone who doesn't know an answer go to another technician he or she thinks will? Now two or more technicians are involved, probably costing your help desk much more to solve the problem than it should.

Even when your tech finds the right solution somewhere, he or she will need to write it down or make a photocopy or carry the manual containing the solution over to a telephone, dial the customer back and explain what's wrong.

And you're likely to end up solving the same problems repeatedly. Once one person in your organization successfully solves a customer's problem there's no reason for someone else to duplicate the efforts. It's costly and completely counterproductive.

A small or new company may get less than 50 calls a week to the help desk early on, but roll-outs of new products, enhancements, etc. are sure to generate more calls. It's best to have an effective operation set up before call volume explodes.

Customers who purchase your products may be paying for support after the first 60 or 90 days. And a paying customer who can't have simple questions answered quickly will not make for a happy customer.

And who can blame them? No one likes to be on hold for long or leave a message and left to wonder when they will receive a call back. It might mean a delay in finishing a project. They may have a plane to catch and

Tips for Buying Tools to Automate

The best tools for problem resolution are those that serve a specific function rather than many. Help desks should be weary of products which promise to be a comprehensive solution. Often, the individual components of the solution are substandard. The exceptions to this are extremely expensive network management and communications packages.

For example, if a quality remote control solution is desired, find the best remote control software package rather than a more comprehensive solution which includes mediocre remote control and many other capabilities. FREE demos to try products out before buying are a great way to make sure the product does what you need it to do. Also help desk managers should make sure the products are easy to use and do not bog the system down with huge memory requirements.

— Avalan Technology

need an answer now. So maybe you decide to give certain customers priority. But that means other customers (perhaps the ones least likely to complain) wait longer to have their problems solved.

Not to mention that if you are not using a knowledge base you are probably relying on well-paid, experienced technicians to solve too many problems. That will drive up your costs.

When you use an automated knowledge base, you can train agents with little technical background how to use some type of artificial intelligence to search your knowledge base for answers. We hear from many help desk who even use temporary workers that are not technically inclined to search knowledge bases using AI. These first level reps solve a very high percentage of calls without escalation.

The results? You're spending less to staff your help desk. Each rep can answer more calls in less time when they use some form of artificial intelligence to search the knowledge base. You'll also spend less time making return phone calls. Customers will be pleased at the quick turnaround time.

Once up and running, many help desks put their knowledge bases on line so their end-users can try to solve their own problems. This is a low-cost way to decrease the number of calls to the help desk.

When making the decision to develop a knowledge base it's important not to under estimate the amount of time and work it will take. "It takes a really good engineer working full-time on building knowledge, about a week just getting data to create 30 to 60 cases," says Ivy Meadors, owner of the consulting firm High Tech/High Touch Solutions (Redmond, WA). "And that doesn't include typing all the information in."

She suggests not having an expert in the application build the knowledge. "This person should be closely involved, but someone who can write at a level for the novice should build the knowledge."

Tips for Buying Problem Resolution Tools

- *Decide on internal vs. external help desk software. Don't try to make one fit into the wrong environment.*

- *Decide on proper support processes before selecting a tool. Have the tool model your process, not your process model the tool.*

- *Pay attention to the ease of building and maintaining an expert systems knowledge base. Cases must be kept current to have any long term value.*

- *A robust SQL database is a must. Can you afford to lose a call ticket if the database on the application fails?*

- *Ease of use in reporting is very important. You must be able to extract data from the system to determine trends that could improve support.*

—Allen Systems Group

Meadors stresses the importance of developing a process for updating the knowledge base as new problems are resolved. She advises against allowing non-technical agents add to the knowledge base. "The knowledge engineer should be responsible for going in and deciding what to add."

She also advises against hiring the vendor from whom you purchase help desk software to build your knowledge base. "That's a very expensive solution," she says. "They'll charge $75 to $100 a hour. It works out a lot cheaper using the average engineer who earns $35,000 to $45,000 a year. You also want to use someone who knows the people in your organization."

Gateway 2000's Knowledge Base Implementation

Over the past year Gateway 2000 has gone through an engineering process. They've reevaluated their procedures for solving problems. They used to use different databases and text retrieval systems, but their three support centers could not share information.

"We had ten years of knowledge that we needed to prioritize," says Gateway's Skip Post. "We needed to decide how to structure the knowledge and what to structure first to give us the biggest return on investment when solving customer problems."

Post says the biggest issue was structuring the information. He says they broke it down into three parts: looking at data which is the raw knowledge; looking at information and assimilating what it means; and looking at the actual knowledge — what they have learned from proven solutions.

"We needed an environment conducive to sharing information. If an agent knows everything that agent will excel," says Post. "But we want everyone to have that knowledge so everyone can excel."

Post says buying the technology was the cheapest thing. The imple-

Tips for Buying Problem Resolution Tools

Our best tip when it comes to buying a problem resolution tool is a simple one: Don't ask if an application performs a specific function, ask how it performs it. In today's help desk application market, our studies show that approximately 75% of all application functionality is the same from one vendor to the next. Thus, to ask whether a product does network management, staff notifications, escalation, or "is tailorable" is not time well spent. Instead, ask how a product performs these functions, and compare them in terms of ease of use, flexibility, and robustness.

Savvy help desk managers should look for the differences from one application to another; also, they should prepare 6-12 site-specific requirements and see how each application addresses them.

By the same token, help desk managers should not judge an application based on some generic functional checklist. Use checklists to decide what functions are important to you, and then use that list to look for potential solutions.

—Applix

mentation is where the real cost is. "Anyone can buy the technology, but you need the right process and infrastructure in place. You need to figure out how to get information, analyze it and get it back to the floor. The idea is to serve customers better and give technicians knowledge at their fingertips."

They purchased a system from Vantive for call tracking and use Inference's system for case-based reasoning. All of their support reps who field calls are seasoned technicians. Over the past few months Gateway put together teams of engineers who have built thousands of cases.

Gateway realizes the importance of ensuring there are no inaccuracies in cases. That's why up front they had many engineers devoted full-time to building cases and several people who maintain the knowledge base. Post says they hire case writers and engineers. The knowledge engineer gets the knowledge and passes it on to the case writer. Once the case writer has written the case in the right style and format the knowledge engineer checks it to make sure it's accurate before it's put online.

Post says they're able to resolve more issues on the first call. "Even if there's no immediate resolution, you should be able to give customers a status and set their expectations."

Post says that in order to successfully build a database, you need to have adequate senior management support and a commitment that you can adequately staff. They plan to put their Inference search tool on the Web soon so callers can use case-based reasoning to search for solutions themselves.

So, does Gateway think investing all this effort in developing a knowledge base was worth it? "We're still in an adolescent stage but we know we'll appreciate the benefits down the road," he says. He also says it's taking time to train and teach staff how this structured system will benefit both them and the customer. But even in the early stages, Post says they've already enhanced their ability to respond to customers.

Artificial Intelligence

Artificial intelligence studies how people think, make decisions and arrive at conclusions. An AI system learns from past experience.

Making an investment into tools that let you search your knowledge base

Smooth Sailing Tips

Make sure the tools provided to the support representatives facilitate problem resolution without injecting frustration. Also, provide a mechanism whereby the support representatives feel they have ownership for customer satisfaction and overall improvement of the product or corporate efficiency.

— Silvon

should be a well-researched project. The first step in determining how any (or several) of these technologies can help you is to weed through all of the AI technologies expert systems use. So here are some explanations of how these technologies work when used in the help desk.

Case-based reasoning — CBR uses past occurrences to find solutions. It will look at how a similar problem was solved and suggest the same solution if the problem seems to match. It can give you a solution that would normally require expertise to figure out, without the need to ask customers many questions. CBR needs a large number of cases in the knowledge base to be effective.

Decision trees — ask a series of questions to which the caller's responses brings (branches) the support rep to another question. Eventually the responses can lead the rep to a likely solution to the problem. This system is most effective when used by reps with little help desk or technical training because the system leads the rep through the process.

Text retrieval — also known as key word searching. These are ways of searching the knowledge base by typing in words that describe the problem in at attempt to come up with the correct solution.

This search tool is best used by experienced reps who are technically inclined and know how to properly describe a problem. Key word searching doesn't use artificial intelligence. A common problem with text retrieval/key word searching is the likelihood that different people will describe a problem differently, making it difficult for the knowledge base to suggest a solution.

Natural language processing — Lets users describe a problem in their own terms to find a solution in the knowledge base. Problems can be entered into the system informally. This technology is most effective, once again, for reps with little help desk or technical experience.

Neural network — learns patterns and relationships in data. The Help Desk Institute (Colorado Springs, CO) defines a neural network as a "hardware and software simulation of the human brain using artificial neurons combined in a massively parallel network."

Fuzzy logic — Can draw a possible solution when there is conflicting information or no exact match for a problem description. Fuzzy Logic assigns a value of confidence based on possible solutions.

Rules-based system — uses logic rules based on the information given. A rules-based system says if X is the problem then Y must be the solution. This system is probably the least common form of AI. One problem is that it requires lot's of pre-programming and is only effective for a small amount of knowledge that isn't likely to change often.

Now that you know what these technologies are, how can you decide what's right for you? First, you need to consider how each of these technologies will or won't benefit you. Each one can be effective in helping you match a problem with a solution. But the degree of effectiveness depends on how you operate your help desk, the size of your operation, the nature of calls and the type of people you hire to solve problems.

Dr. Lance Eliot, president of the consulting firm Eliot & Associates (Huntington

Tips for Buying Problem Resolution Tools

1. Flexibility: Is it easy to look around the database, moving forward and backward in your search process to find the answer?

2. Quality of knowledge it works with. The top two vendors in the knowledge business are KBI and ServiceWare. Make sure that your system works with one of these two vendors or you will be in the knowledge development business whether you like it or not.

3. What kinds of people are going to be using your problem resolution system? For highly trained analysts, a key word search is often the most helpful, because they will know exactly what they're looking for, and a fancy search engine only slows them down. For a low level analyst, a more intuitive question-and-answer session may be needed, because they're not really sure what the answer to the problem is, so the problem resolution process is more than a simple search for them. If you have a mixture of these two types of people, then you should have a mixture of these types of systems available in your help desk product.

4. Are you likely to need to build your own knowledge base into the sophisticated problem resolution tool? Then dig deep on what is really involved in building the trees, cases, text bases, what have you, because this could be a major cost and time issue for you, as the ease with which you can build this knowledge is critical to whether it ever actually gets built, let alone used.

5. Is the knowledge you need to build up time-sensitive, will it change every few months? If so, seriously question any type of knowledge building that is time intensive. With a great deal of change taking place, the time needed to write the knowledge base won't really be worth it. It is better to keep track of it another way, i.e., by logging the resolution in the help desk product (in the process of logging the call, which you would have to do anyway) and making it easy to find.

6. Does the system have to be customized to work for you? This is a big cost issue, so make sure you discuss this thoroughly with any vendor. Come upgrade time, if you have customized your product significantly, you may be looking at a big hassle and a big bill.

7. Is the help desk product and the problem resolution product integration meaningful? For example, if you fire up the problem resolution tool, does it start the search based on information you already entered in logging the call? If not, this kind of integration is available in other tools, so look around. This is really helpful in speeding the process, and putting the analyst on the right track in solving a problem.

8. Does it work in your environment? A product is not going to be much help if you have to change your environment around to make it work. Get what works with what you already have or are already planning to move to.

9. Are they selling you on self-help as a call-avoidance method for your internal help desk? If you are looking for a customer support solution where you have a finite number of possibilities, this MAY be an option, but for internal support it isn't.

—Utopia Technology Partners, Inc.

Beach, CA) says it's important to clarify the purpose of AI. "AI strives to make the computer more intelligent, imitating the intelligence of a human being," says Eliot. "But there's a difference between the goals of artificial intelligence and what we can with it do today. The goals come nowhere close."

Still he thinks the technology can be very useful. "The sad thing in the help desk is that people buy an automation tool that contains an AI component but never use it because of the complexity," says Eliot. "They make the investment, but it becomes too difficult to figure out or they can't figure out how to apply it."

He goes on: "A classic problem help desk managers face is upper management telling them to get some kind of artificial intelligence up and running, but if that person still has to field calls or is just more socially oriented, they will spend the most time at the task they're most comfortable with."

The artificial intelligence we see today will look different in the future. Many of today's AI technologies try to meet human needs but don't necessarily have the same sense as a human being.

Some of the technologies cannot always scale down to the problem's solution as effortlessly as some vendors might make it seem. Since using expert systems in the help desk is relatively new, it should not be long before we start to see improved, faster and easier to use products.

Rushing to select an expert system is a big mistake help desks make. It is vitally important that once you decide how this system can benefit you, you develop a selection process. Eliot recommends doing a cost benefit analysis figuring out the costs of following through each call without searching a knowledge base vs. digging up a solution once and then always being able to refer back to it.

"When evaluating tools, help desks make the mistake of checking off all the different technologies they think they'll need without really thinking about how they will use them," says Eliot.

Some help desks employ support reps with little technical experience while other help desks only use well-trained technicians to handle all calls. Other help desks use relatively non-technical support reps at the first level of support and use technicians with technical savvy to handle problems that can't be solved at level one. The type of people you use on your help desk affects the type of product you should use.

If someone is a seasoned technician it can be frustrating for them to be forced to key in whole sentences to take advantage of a natural language interface when trying to search the knowledge base.

"If the system doesn't let them take the quicker route by just using key words to search the knowledge base, then it will create frustration," says Eliot.

On the other hand, if your help desk uses mostly non-technically oriented reps it can be frustrating to them if they can only search the database using key words. A problem with simple key word search is that different people will use different key words making it difficult to search the knowledge base if they don't hit on the right word to bring up a viable solution.

In this case a natural language interface would be more effective. The user can enter complete sentences and the computer will comprehend what they are trying to do. You will find some products that claim to use a combination of different AI technologies. The claim that the product really offers all of these distinct techniques to search the knowledge base may be true. But you also need to ensure how well the vendors have implemented these techniques.

"The range may be limited," says Eliot. "There may be some kind of neural network or fuzzy logic but it may be a reduced version. It's important to make sure the product's claims are bona fide. Vendors want to equal up to a help desk manager's check list with all the features they think they need, so in order to stay in the game, they may feel they have to claim their product utilizes many of these technologies."

Eliot suggests installing the product on a small scale and working with it. That's the only way to really see if it can meet your needs. And you need to have someone skilled enough to do this on staff or else hire a consultant.

"There's a real process you need to go through," says Eliot. "It's not enough to pick a product based on a simple demo or spending some time at a vendor's booth during a trade show. You need to evaluate your needs now and six months or a year down the road. I've seen help desks rush the process of selecting a problem management tool only to find they're stuck with a product that lacks the features they need after they've made a large investment.

Tips for Buying Problem Resolution Tools

Beware of proprietary solutions ... this means products that are out-of-the box. Since users reconfigure processes to fit the product's configuration (workflow etc...), often it arises that at least part of a changed process needs to be retained. This means going back and trying to customize a solution that isn't designed to be customized. It also serves as a problem when processes change. Although an out of the box solution works for today's needs, those needs do - and most likely will - change. This also means trying to configure a solution that wasn't designed for customization.

Also beware of significant nickel and dime charges ... companies may purport low entry fees, but then rack up charges with consulting or major licensing fees.

Most of all, it is imperative that users understand and know the processes that must be changed, and find a solution that not only provides an answer for today, but also allows for changes to fit with evolving needs.

— Remedy

➲ Chapter 4: Where To Get Problem Resolution Tools

This is the mother of all guides listing just about every vendor that makes problem resolution/help desk software or other system for use in the help desk.

Advantage kbs, Inc.

Corporate Headquarters:
1 Ethel Road, Suite 106B
Edison, NJ 08817
Phone:
908-287-2236
800-AKBS-YES
Fax:
908-287-3193
E-mail:
info@akbs.com

Products:
The IQSupport Application Suite
IQSupport Pro
IQSupport Lite IQAdvisor
IQWeb (Release date 12/96) Knowledge Editor

Suggested List Prices:
IQSupport Pro: . . . .$750/concurrent user
IQSupport Lite: . . . . $50/concurrent user
IQAdvisor: . . . . .Royalty arrangement
IQWeb: . . . . .Price to be determined
Knowledge Editor: $7500/copy

Compatible with (network): NetWare, Digital Pathworks, Banyan VINES, Artisoft LANtastic, Microsoft LAN Manager, IBM LAN Server

Compatible with (platform): MS Windows 3.1, MS Windows 95, MS Windows NT, HP-UX

Product Description:

The IQSupport Application Suite is designed to provide customer support representatives, help desk staff, field service technicians, and customers with all the information they need to solve problems. IQSupport is problem resolution software specifically designed to aid novices in solving problems.

IQSupport now has a multi-browser interface which allows electronic manuals to be viewed using WinHelp, DynaText, Netscape Navigator, MS Word and Notepad browsers.

IQSupport Pro includes APIs for integration to call tracking systems, billing, database and other applications; is integrated with the Enterprise System from Vantive, and the Action Request System from Remedy; and uses ServiceWare Knowledge-Paks.

IQSupport Lite empowers corporate end-users to solve problems independently and uses ServiceWare's Knowledge-Paks.

IQWeb provides the same troubleshooting capabilities to customers and staff visiting World Wide Web sites.

IQAdvisor allows organizations to embed this powerful problem resolution engine into their help system.

The Knowledge Editor is the authoring tool to create custom knowledge bases for all IQSupport products.

Allen Systems Group

Corporate Headquarters:
750 11th Street South
Naples, FL 34102

Phone:
941-435-2200
800-932-5536

Fax:
941-263-3692

Web:
www.allensysgroup.com

Product:
ASG IMPACT

Suggested List Price:
$62,000 and up depending on product options chosen

Network Compatibility: Novell LAN, IBM, LAN Server

Platform Compatibility: IBM Mainframe, 0S/2, Windows

Product Description:
ASG-IMPACT is a high-lvel customer support tool that aligns an organization's IT with their lines of business through service desk consolidation. ASG-IMPACT offers Enterprise Service Management by combining the disciplines of service of existing mainframe, as well as client / server configurations. The heart of the system is a dynamic event management hub called Distributed Service Facility, that acts as a collaborator, integrating information from system management facilities into the consolidated service desk.

The system also offers an integrated knowledge base system called ASG-IMPACT/Expert. This feature provides the ability to create company-specific knowledge bases as well as importing knowledge from ServiceWare.

Amtelco

Corporate Headquarters:
4800 Curtin Drive
McFarland, WI 53558

Phone:
800-356-9148
608-838-4194

Fax:
608-838-8367

Web:
www.Amtelcom.com

Product:
CallScriptor, The Enhanced Messaging System

Suggested List Price:
$17,000+ depending on # of stations

Product Description:
CallScriptor lets the user easily create custom scripts with sophisticated branching capabilities. Help desk is one of many applications CallScriptor can handle. CallScriptor is a Windows application operating on a Novell LAN. CallScriptor uses Microsoft Access as the main Reporting module and is written in FoxPro. Past history is easily obtainable using the FoxPro DBF files. CallScriptor is also used in telemarketing applications.

Tickets can be opened and tracked until resolution and information can be easily dispatched through different types of media. History can be gathered and reported on in many different types of formats. Decisions can be made quickly and easily based on the different types of inquiries reported. CallScriptor has the capability for pop-up screens using Windows DDE links.

CallScriptor is also designed to handle other types of inbound calls requiring database access, customized scripting, order entry, survey and brochure requests.

Applix, Inc.

Corporate Headquarters:
112 Turnpike Road
Westboro, MA 01581

Phone:
508-870-0300

Fax:
508-366-2278

E-mail:
applixinfo@applix.com

Web:
www.applix.com

Products:
Applix Enterprise is a suite of interactive customer service applications, consisting of the following: Applix Helpdesk; Additional products and services for Applix Helpdesk; Executive Information Center; SupportLink; Application Design; Form Editor; LifeLine MobileLink; LifeLine/HA; Applix Service; Additional products and services for Applix Service; Applix WebLink

Compatible with (network): Novell NetWare Banyan VINES LAN Manager Microsoft Windows for Workgroups Pathworks TCP/IP

Compatible with (platform): Windows 3.1 or 3.11 Windows 95, Windows NT, UNIX, OS/2

Product Description:
Applix Enterprise is a suite of integrated software applications which automate and improve customer service and support processes, regardless of an organization's size or discipline. All Applix Enterprise applications and modules tightly integrate with each other to provide an enterprise-wide solution. Applix Enterprise software is also designed to be compatible with leading packages for network management, automated asset collection, notification technologies, E-mail and more.

Applix Enterprise line includes:
Applix Helpdesk - Automates and improves internal support operations through tailorable workflow and resolution technologies.

Applix Service - Provides a comprehensive solution to support customers and use their feedback to improve the quality of an organization's own products and services.

Interfaces of Applix Enterprise:
Applix WebLink - Allows real-time access to an organization's support center to create a virtual service environment that meets customers' demands more effectively.

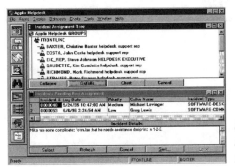

Shown are the Incident Assignment and Rep Workload Windows in Applix's help desk.

Applix SupportLink - Gives customers (internal and/or external) direct computer access to an organization's support center to log, research, and resolve questions and problems on their own.

Executive Information Center - Puts Applix Enterprise charts and reports on the desktops of managers and executives, providing them with a concise understanding of an organization's operations.

Applix LifeLine/HA - Guarantees the operation of an organization's support center even in the event of a network or server failure.

Applix Mobile - Allows information to be processed and updated without direct connection to the central database.

Avalan Technology, Inc.

Corporate Headquarters:
7 October Hill Rd.
P.O. Box 6888
Holliston, MA 01746

Phone:
800-441-2281
508-429-6482

Fax:
508-429-3179

E-mail:
saleinfo@AVALAN.com

Web:
www.AVALAN.com

Products:
Remotely Possible Suite

Product Description:
Remotely Possible is remote control Windows PC software used to access, take over and monitor a remote PC to offer support, to train, to do stealth monitoring of PC's, to transfer files, to chat, to do remote printing and more.

In addition to remote control, all Remotely Possible products include:

- Security – Multiple security features include host call-back, user ID and password.
- B7 Printer Redirection – This feature enables you to print at the Host, viewer or both.
- MultiHost – Connect to more than one remote site at the same time.
- Role Reversal – Switch between Host and Viewer mode without having to reestablish the connection.
- Reboot Remote PC – Restarts the remote system to load updated drivers and/or system files.
- TCP/IP support for support over the Internet and RAS.
- TAPI support for direct dial-in support.
- File Transfer – Upload new driver/program files to troubled PC. Download system files or other information for troubleshooting purposes.
- Chat – Real time two way keyboard chat to assist in communicating to resolve the problem.

Family of Products:

Remotely Possible/LAN — 16-bit Multi-Server Novell, IPX, NetBios & Windows for Workgroups Networking version. $599 up to 100 users at one site. Multi-site packs available.

Remotely Possible/ Dial — 16-bit Dial up remote control modem version for travel, home-office, laptop, & dial-in, dial -out connections. Licensed for two PCs. $169 (license for two PCs) Quantity discounts/site licensing available.

Remotely Possible/Sockets — 16-bit support for TCP/IP via Windows Sockets. Perfect for the Internet, PPP, SLIP & Dial In connections. Advanced compression, stealth remote access for discreet viewing. $169 (license for 2 PC's).

Remotely Possible/32 for 95 — 32-bit version for Windows 95 users. Supports multiple Hosts & Viewers, Passive Monitoring, five level security, simultaneous remote control, file transfer & chat, data encryption, Group Viewing, for Help Desk, Call Centers, support and training. Offers on-line help, takes only 2Mb hard disk. Connect via Dial up, TCP/IP, RAS, PPP, SLIP, & TAPI. $169 (license for two PCs).

Remotely Possible/ 32 LITE — Free 32-bit version from Web to try the fully functional version for 30 days. After 30 days, version becomes lite. Keep the Lite version forever or upgrade to RP/32. Free via www.Avalan.com.

Remotely Possible/32 for NT — Control any Windows NT Workstation and NT Server from any Windows Operating system. Simultaneous remote control, file transfer, keyboard chat, & Passive monitoring. Connect over Winsock TCP/IP, PPP, SLIP, RAS,TAPI, & IPX. Unique multi-takeover features ideal for Help Desk, Call Centers, classroom/training and support. $199 for workstation pack. $399 server pack.

Tips for Buying Problem Resolution Tools

Choose the right vendor and carefully plan your implementation. A carefully planned, strategic implementation is the key to a successful help desk. Implementing a successful help desk system is 50% planning and selecting the right vendor product, and 50% implementation. The implementation process should include the following phases and steps to accurately reflect the nature of the help desk: define your requirements, assess the products and vendors, formulate the project plan, install the software and load the database, test the system, train the staff, run the pilot program, and deploy and refine the system. For step by step instructions, refer to Datawatch's white paper entitled: Implementing Your Help Desk: A Practical Guide.

— Datawatch

Baron Software Services

Corporate Headquarters:

584 Ardsley Blvd.
Garden City, NY 11530

Phone:

516-292-2323

Fax:

516-292-2385

Product:

Manage-it

Suggested List Price:

$249 for complete package with software and unlimited phone assistance. (multi-discounts are available).

Compatible with (platform): All Windows versions and OS/2

Compatible with (network): NT Server, Novell Netware, Warp Server & most peer-to-peer networks

Product Description:

Help desk software for call tracking, with data collection and reporting tools, including the Know-it expert knowledge base system. The system lets you set priority levels and do auto request numbering.

Bendata, Inc., an Astea International Company

Bendata Adress:
1125 Kelly Johnson Blvd.
Colorado Springs, CO 80920

Corporate Headquarters:
Astea International Inc.
100 Highpoint Drive
Chalfont, PA 18914

Phone:
800-776-7889
719-531-5007

Fax:
719-536-9623

E-mail:
sales@bendata.com

Products:
First Level Support (Bendata); HEAT Premiere (Bendata); HEAT for Windows Professional Edition (Bendata); CASE-1 (Astea); PowerHelp (Astea)

Suggested List price:
First Level Support . . . .$95 and up
HEAT Premiere . . . .$1,195 and up
HEAT for Windows Professional Edition . . . .$1,600 and up
CASE-1 . . . .$5,000 and up
PowerHelp . . . .$4,000 and up

Compatible with (network): Any network that has drive mapping and DOS 3.x file locking capabilities (HEAT series). Windows, UNIX, Windows NT, VMS (PowerHelp)

Compatible with (platform): Windows 3.x and higher, Windows NT, 8 MB RAM, 40 MB disk space (HEAT series) Windows 3.x and higher, Windows NT, 16 MB min., 62 MB disk space (PowerHelp)

Product Description:
Both Heat and First Level Support offer a "tree" structure for narrowing down problem solutions. First Level Support offers knowledge based problem solving and is a part of Heat for Windows Professional Edition.

Heat Premiere is designed for start-up or small help desks with low call volume (up to five users). The system offers a Windows platform, 40 predefined reports.

Heat for Windows automatically logs calls and generates trouble tickets and offers escalation features. The statistics monitor can display user-defined information.

CASE-1 is Astea's case-based reasoning system which helps service organizations diagnose a situation and suggest viable solutions to solve end-user problems. The system employs adaptive learning, adjusting to the user's environment. New solutions can be created by combining prior experiences that share common elements. As a result, the system continually "learns" new solutions and becomes increasingly more "intelligent."

PowerHelp is their client/server help desk automation system (which is fully integrated with CASE-1) that enables support reps to capture, analyze, route and resolve customer calls in both high volume internal help desks and external customer support centers.

First Level Support from Bendata (owned by Astea International) lets you access files to solve a problem

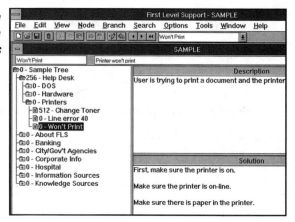

Clarify Inc.

Corporate Headquarters:
2702 Orchard Parkway
San Jose, CA 95134

Phone:
408-428-2000

E-mail:
Info@clarify.com

Web:
www.clarify.com

Products:
The Clarify Product Family
ClearSupport; ClearQuality; Clear Helpdesk; Clear Logistics; ClearSales

Pricing:
Priced at $1,000 to $3,750 per user based on solution configuration, the Clarify system features two core application options: ClearSupport for external customer support environments, and ClearHelpdesk for internal help desk management environments.

The Clarify system runs on standard client, server and database platforms. Servers supported include Sun, Silicon Graphics, IBM, and HP for UNIX; and Intel x86 and Pentium for Windows NT. Clients supported include Windows, UNIX/Motif and Macintosh; databases supported include Oracle, Sybase, and SQL Server.

Product Description:
ClearHelpdesk is a complete solution for handling employee questions or suggestions about technology, benefits and facilities. The system interfaces with popular network and system management tools, and off-the-shelf knowledge bases to provide benefits like employee call tracking asset management, change management, service level agreements, problem diagnosis, workflow management and report generation.

Clarify's comprehensive product suite includes sales and marketing, customer service, problem resolution, field service, logistics, defect tracking and help desk applications. The products handle all facets of customer interaction over LANs, WANs, Internet and mobile connections.

This is a service level agreement screen part of Clarify's ClearSupport

Clientele Software

Corporate Headquarters:
8100 S.W. Nyberg Road
Tualatin, OR 97062
Phone:
503-612-2600
Fax:
503-612-2800
Products & Pricing:

Clientele /IS: . . . . $8,695 (5 users)
. . . . $11,995 (10 users)
. . . . $18,495 (20 users)
. . . . $36,995 (50 users)
. . . . $64,995 (100 users)
Clientele:. . . . . same pricing as above
Clientele Conductor . . . . $2,595 per database
Client/Server option . . . . $5,995 per database
Replication Server . . . . $1,295 per site
. . . . $895 single remote user

Product Description:

Clientele is customer service software for Microsoft Windows and Windows 95 and NT Environments that automates all aspects of external customer service including order entry processes.

Clientele/IS shares the same functionality but also adds features like asset tracking and inventory management. The system features AnswerBook, their problem resolution facility, and supports 32-bit and 16 bit environments. Problem can be described in natural language and these descriptions are sent to a search engine that retrieves possible solutions from collections of self-authored database, prepackaged knowledge or previous call records. The search engine technology is licensed from Verity. The system is also available for DOS environments.

Conductor is an optional module for monitoring open calls and automatically taking action, such as reassigning calls, issuing follow-ups, running reports, activating pagers, etc.

Replication Server lets users share a common set of data at different sites. Users at each site work with a local copy of their data that's periodically updated via modem or WAN connection from other Clientele installations.

ConsulNet Computing

Corporate Headquarters:
51 Paperbirch Dr.
Don Mills, Ontario, Canada M3C2E6

Phone:
416-441-0363
800-361-9527

Fax:
416-449-1985

Product:
ConsulNet Support Log

Suggested List Price:
$599 single user,
$1,499 five user,
$2,499 unlimited users

Compatible with (network): DOS based network, Novell, Lantastic

Compatible with (platform): Windows 3.x or higher, Windows 95 & NT

Product Description:
ConsultNet Support Log tracks the status of open and closed calls by user, division, department, staff assignment, charge to and date. It can be used for help desk calls, software maintenance, customer support, service repair tracking, dispatch logging, etc.

Datawatch Corporation

Corporate Headquarters:
234 Ballardvale Street
Wilmington, MA 01887-1032

Phone:
508-988-9700
800-988-4739

Fax:
508-988-0697

Web:
www.datawatch.com

Product Name:
Q-Support (TM)

Pricing:
$6,500 (minimum 3-user license)

Compatible with (network): Runs on any LAN system that supports DOS record locking 3.3 or above

Compatible with (platform): Microsoft Windows v3.1 or later

Product Description:

Q-Support is a comprehensive help desk automation and asset management system. The program provides IS support centers with a wide range of PC software tools for use in a multi-user, networked environment.

Featuring a graphical user interface, Q-Support logs, routes and tracks calls from initiation through resolution. Q-Support also provides inventory control, the ability to monitor performance levels, access to third-party knowledge tools, multiple Service Level Agreements (SLAs) and extensive reporting options. SMTP, VIM and MAPI links enable the help desk to receive and respond to calls via e-mail.

Q-Support's client/server version is multi-platform; the software supports many popular data server engines such as Microsoft SQL Server and Oracle 7.

DCS Software & Services

Corporate Headquarters:
2610 Highway 544 West
Wylie, TX 75098

Phone:
214-429-8200
800-999-2495

Fax:
214-429-8216

E-mail:
jim@dcscorp.win.net

Product:
Problem Managementplus

Suggested List Price:
$3,995 / site license

Compatible with (platform): IBM AS/400

Product Description:
Problem Managementplus is a tracking system designed specifically for the IBM AS/400. The system is user configurable to your environment and offers features like user defined reports, automatic escalation monitor, and the ability to interface with outside databases and programs.

EXSYS, Inc.

Corporate Headquarters:
1720 Louisiana Blvd. NE Suite 312
Albuquerque, NM 87110

Phone:
505-256-8356

Fax:
505-256-8359

E-mail:
info@exsysinfo.com

Web:
www.exsysinfo.com

Products & Prices:
EXSYS RuleBook Cost..$1,495.
EXSYS Professional Cost...$2,900.
EXSYS Linkable Object Modules Cost ....$5,000.
EXSYS Web Runtime Engine (WREN) Cost ....$7,500.

Compatible with (network): Novell, Banyan, Ethernet, TokenRing, more

Compatible with (platform): DOS, Windows, Windows NT, Windows 95, Macintosh, OS/2, VMS, Sun Solaris, Sun SunOS, SCO UNIX, BSD UNIX, HP-UX, HP, more.

Product Description:
Software tools for building and disseminating problem-solving knowledge-based expert systems. Proactive help desk systems automate answers to the most frequently asked questions. Answers to questions are available 24 hours a day, automatically, on a Web page or disseminated on disk to your end users. Systems are developed by your in-house domain experts. Applications explain to end-users how conclusions were reached.

EXSYS Professional: For developing probabilistic, knowledge-based expert systems using IF-THEN-ELSE production rules. Generic hooks allow invisible embedding and easy interfacing to other applications, SQL (ODBC) databases, spreadsheets, neural nets and process control software. Help screens select items from lists of various options for assigning and combining probability values. The Editor, Runtime and Utility programs are written in "C".

Command Language uses sub-sets, looping and conditional tests; design program for customization of screens; automatic validation and report generation. Includes unlimited runtime license. Portable across platforms. Hypertext, frames, fuzzy logic, security, six confidence modes, LAN and site licenses.

EXSYSRuleBook- Tree diagram style expert system development software. Systems are built using tree diagrams that describe an entire aspect of a problem. Nodes in the tree represent questions asked of the end user. When a new node is added, other branches are automatically built for all possible input values. Multiple trees are created representing independent aspects of a problem. Backward chaining combines the trees into a single system. "Balloon help" windows show rules associated with branches.

EXSYS Professional Linkable Object Modules- For extensive customization. Allows calling the EXSYS Professional inference engine as a "C" routine from within another program. Add up to 100 user defined functions, which behave like internal EXSYS commands, to perform special functions for data acquisition or display. Overlay EXSYS Professional with other programs which reduces the memory required to run large programs.

EXSYS Web Runtime Engine (WREN)- Expert systems developed with EXSYS tools makes decision-making knowledge and logic available on the Web. Users are asked questions by the system which then recommends the best solutions to problems. Multiple users can simultaneously access expert knowledge for tech support, product selection, help desks, regulatory compliance, etc. Provides on-line expert knowledge 24 hours a day. It also allows secured knowledge from one site to be accessed and shared by others via an intranet.

Fujitsu Software Corporation

Corporate Headquarters:
3055 Orchard Drive
San Jose, CA 95134

Phone:
800-446-4736
203-326-2700

Fax:
203-964-1007

Web:
www.fsc.fujitsu.com

Product:
LiveHelp

Suggested List Price:
Expert: . . . .$99.00 per Expert Client:
. . . .$175.00 for 10 Pack

NetWork/Modem Support: Novell/IPX, TCP/IP, Modem Support 14,400 or greater

Platform Compatibility: Win 95, Win 3.1, NT

Product Description:
LiveHelp is a Windows remote support product that provides screen-sharing and remote-control capability so that service departments can offer computer users real-time, on-line configuration system support and training.

Support personnel can provide solutions to users by teaching them problem resolutions to reduce future repeat calls. Multiple experts can conference at one time allowing more experienced experts to help first-line technicians.

LiveHelp's training sessions, whiteboard and file-transfer features allow help desk personnel to resolve technical problems.

Smooth Sailing Tips

• Don't pass responsibility for problem solving. • Bring other experts into the process. • Establish ongoing first level support training. • Develop service level agreements to establish expectations. • Never count on one person to support a product. Be sure to cross train your support personnel. • Rotating and training staff will help retain employees and improve customer satisfaction.

— Allen Systems Group

IBS Corporation

Corporate Headquarters:
12626 High Bluff Dr.
San Diego, CA 92130

Phone:
619-792-0273
800-346-2894

Fax:
619-792-5199

E-mail:
ibs@cts.com

Products:
CONFerence-PC CONFerence-400 CONFerence-HOST EvaluAtor-PC
EvaluAtor-400 EvaluAtor-HOST

List Price:
CONFerence-PC. . .$250 to $15 per Workstation depending on volume.
CONFerence-400 . . . .$2,500 to $1,000 per AS/400
CONFerence-HOST . . . .$100,000 to $25,000 per host
EvaluAtor-PC . . . . .$150 & up per workstation, depending on volume.
EvaluAtor-400 . . . .$2,000 to $500 per AS/400
EvaluAtor-HOST . . . .$50,000 to $10,000 per host

Compatible (network): PC: IPX, TCP/IP, Banyan Vines, Asynchronous, NetBIOS, X.25, point-to-point, Internet. AS/400: any AS/400 communication protocol. HOST: any 3270 or compatible protocol.

Compatible (platform): PC: DOS, Windows 3.1, Windows 3.1.1, Windows 95, Windows NT; AS/400: Any operating system; HOST: CICS

Product Description:
(FIX) Allows users to see and capture user problems rather than guess, and while they are seeing and capturing the users problems they can pop a customized logging form or their own in-house tracking system or any other tracking system. Also allows an authorized person to take over, control or perform a cooperative control of the user's workstation to resolve the problem. The CONFerence technology can also be set to capture/communicate/dispatch to a support person based on events that may occur on the user's workstation.

Inference Corporation

Corporate Headquarters:
100 Rowland Way
Novato, CA 94945

Phone:
800-322-9923
415-893-7200

Fax:
415-899-9080

E-mail:
info@inference.com

Web:
www.inference.com

Products:
CBR2 Family of Products, which includes: CasePoint; CasePoint
Search Engine; CasePoint WebServer; CBR Express; CBR Express
Tester; CBR Express Generator

Suggested List Price:
$1,000 to $2,500 per seat, depending on configuration.

Compatible with (platform): Windows 3.1, Windows 95, Windows
NT, Sun Solaris, HP-UX, OS/2

CBR2 Database Support for: Oracle, Sybase, Microsoft SQL
Server, Informix, DB2/2, RAIMA Data Manager (RDM)

Product Description:
CBR Express is the primary knowledge authoring tool of the CBR2 family
of products. It uses an intuitive, fill-in-the-form interface that allows direct
natural language input by domain experts to build case bases for problem
resolution or product selection.

The solutions CBR Express creates (incorporating text, graphics, anima-
tion, and sound) are easy to use, but made possible through the advanced
CBR technology kernel that drives CBR2.

Authors can modify every component-including the types of questions, the
"weight" of answers, the integration of multimedia, and more. They can
add rules that help users make better choices or trigger external routines
that correct online problems automatically.

Since CBR Express builds solutions incrementally, you can put them online
as soon as your first case is finished. New cases can be added to the case
base dynamically by authors or by the users themselves using the same
application, ensuring that CBR2 solutions improve with use. CBR Express
is available as a Windows application.

CBR Express Generator takes any document and creates an immediately usable case base. It does this by statistically analyzing the content of the corpus it is given. It automatically provides a summary based on those phrases which are statistically relevant within each document, and questions which help to differentiate between similar documents. You can then use the conversational search capabilities of CasePoint, rather than Boolean searches often required in text retrieval systems. The source documents given to the Generator can come from reports, manuals, sales procedures, product specs, technical notes, etc., in fact whatever exists in ASCII, Word Processing, or HTML formats. The CBR Express Generator is available as a Windows application.

CBR Express Tester validates your case base to ensure accuracy and consistency and simulates usage patterns to spot redundancy and verify search reliability. The Tester lets you manage your case base and work without worrying about anything except customer satisfaction. CBR Express Tester is available as a Windows and OS/2 application.

Search and Retrieval Tools: CasePoint CasePoint is the CBR2 out of the box viewer application, that can be used to access the information held within the case base quickly. CasePoint is designed to be a small footprint application.

Type in a query or the symptoms of a problem and CasePoint immediately searches its knowledge base of known information (online documents and examples-or cases-of previous experience). In fractions of a second, CasePoint presents a list of related questions to help narrow the search and ensure that the final result is the closest match to your requirements. This interactive prompt-and-response technology can make novices seem like experts with little or no training, and can provide experts with fast access to information they may know exists, but are unsure where to locate it. CasePoint is available as a Windows, OS/2, HP-UX and Sun Solaris application.

CasePoint Search Engine provides access to all the functionality of CasePoint through a comprehensive Application Programming Interface (API). Developers can provide the benefits of case-based retrieval, in any application of their choice, for example to provide problem resolution, policy or procedure guidance, product recommendation, automatic diagnosis or knowledge publishing capabilities.

CasePoint WebServer provides CBR2 search and retrieval facilities using most standard World Wide Web (WWW) browsers. CasePoint WebServer is an interactive server-based Web application that allows users to access the answers or information they need through the Internet or an intranet. CasePoint WebServer supports the following:

• HTML to provide the interface to the case base • screen customization to adhere to a corporate web style • centralized content management • worldwide access to up-to-date information that requires no additional software by the end user.

Magic Solutions, Inc.

Corporate Headquarters:
10 Forest Ave.,
Paramus, NJ 07652

Phone:
201-587-1515
800-966-9695

Fax:
201-587-8005

E-mail:
info@MagicRx.com

Products:
SupportMagic, SupportMagic SQL

Suggested list price:
SupportMagic . . . .$2,995 for one user
SupportMagic SQL . . . . $5,995 for one user

Compatible w/network: SupportMagic - Network Protocol: NetBIOS,
IPX/SPX (Novell), TCP/IP SupportMagic SQL - Networks: NetBIOS,
IPX/SPX (Novell), TCP/IP

Compatible w/platform: SupportMagic - Network Operating System:
Novell Netware, Banyan Vines, Microsoft LAN Manager, IBM OS/2 LAN
Server, Windows NT and others. Client Operating System: Windows 3.1,
Windows for Workgroups 3.11, Windows 95, Windows NT 3.51 or later
SupportMagic SQL

Server Operating System: Microsoft Windows NT Advanced Server
3.51 or later; Client Operating Systems: Windows 3.1, Windows for
Workgroups 3.11, Windows 95, Windows NT 3.51 or later, OS/2 version 3

Product Description:
SupportMagic: SupportMagic 3.40 is designed for help desks that require
minimum database administration. SupportMagic runs on any popular LAN
from an existing server, with unlimited support for different help desk
groups, and a workflow optimized for internal support. Users can customize fields and screens by support staff, support group or system-wide.

SupportMagic is a native Windows MDI application with a standard problem solution database, complete inventory and configuration management,
and an integrated API for VIM and MAPI e-mail packages. System includes
Magic Solution's own, full-text search engine, external knowledge bases
from ServiceWare, Crystal Reports Professional, WinBEEP paging software, and Magic University training.

SupportMagic SQL: SupportMagic SQL converts the application architecture from file server to client server, for organizations that want to consolidate and manage multiple help desks within a single database. Users can define new tables, fields, links and workflow. Multi-level escalation schemes provide alarms and graphs that are updated in real time. Magic's own search engine retrieves answers from SupportMagic SQL and external knowledge bases in any format; also searches client's e-mail messages and sends back answers to problems. The SupportMagic SQL system supports Watcom, Sybase SQL Anywhere, Microsoft SQL Server and Oracle. The WebMagic module links Netscape Navigator and Microsoft Explorer browsers to SupportMagic SQL. Complete computer telephony and mainframe integration is available.

3. SIR Search Engine (Statistical Information Retrieval): The heart of SupportMagic's problem management and resolution strategy is Magic's own full-text search and retrieval engine. Using embedded artificial intelligence and neural networks, SIR automatically builds your own "experience base" as calls are entered and resolved. Intuitive searching and high-speed performance results in fast, accurate answers to even short-term, repetitive incidents — without any development on the part of users. Using SIR, users can solve problems in "plain English," without rigid, time consuming data entry.

The Word Navigator presents analysts with related words to fine-tune the search. The Self-Help Desk feature enables clients to use e-mail and the World Wide Web to create their own call tickets, quickly get answers to common problems, and even access SupportMagic directly. SIR is seamlessly integrated with the SupportMagic database, company data, and external knowledge bases.

McAfee

Corporate Headquarters:
2710 Walsh Ave.
Santa Clara, CA 95051
Phone:
800-332-9966
Fax:
408-970-9727
Web:
www.McAfee.com

Products & suggested list price:
DP Umbrella SQL v3.10 . . . .$ 9,500 /10 user
VycorEnterprise v3.10 . . . .$45,000 /10 user
VycorWeb v3.10 . . . . $25,000 /site license

Compatible w/network: Novell, Microsoft NT

Compatible w/ platform: Client, Windows 3.x, Windows NT, Windows 95

Database: Microsoft SQL Server NT, Sybase System 10/11 NT, Oracle NT

Product Description:
DP Umbrella SQL v3.10 has three components: A client/server database application used for call tracking and asset management. The database structure provides the repository for network, component, personnel, activity information and much more. The DP Umbrella SQL client is a fully Windows compliant interface and provides easy access and manipulation of the data stored in the repository. DP Umbrella SQL maintains integrated tables for systems, components, personnel activities, and tasks.

DP Umbrella Database Administrator: DP Umbrella Database Administrator is an administration tool used to install, upgrade, and maintain the DP Umbrella SQL database. The Administrator can also import data into and export data from the DP Umbrella SQL database.

DPU Orchestrator: DPU Orchestrator is a server-based transaction processor that acts as an automated attendant for the DP Umbrella SQL database. This application creates and updates activities, sends notifications via pager or e-mail, escalates activities, retrieves e-mail, and performs transaction based processing.

Vycor Beacon: One of the problem prevention features of VycorEnterprise. Vycor Beacon enables the help desk to query the database for individuals based on their equipment, software, location, department, etc.. and compiles a list of recipients. This targeted list is then sent a message informing them of issues that may affect their work day and possible workarounds. This prevention strategy helps get information out to customers before they discover a problem and there by reduces the number of phone calls into the help desk.

Vycor SNMP Listener: This application listens for SNMP (Simple Network Management Protocol) alerts from network management consoles and is another key component in the problem prevention strategy of VycorEnterprise. When alerts are received, the SNMP listener creates a white board notice or an activity ticket. The appropriate technician(s) are then notified of the incident via an e-mail or page. This automatic generation of a ticket provides the help desk with an early warning before customers start to experience problems.

Vycor SMTP Listener: The SMTP (Simple Mail Transfer Protocol) Listener is capable of receiving and carrying out instructions from other network applications that use SMTP. This application can create new tickets and whiteboard notices as well as update information on existing tickets such as priority and status.

Rescue Rescue is a desktop application that is used by help desk customers as an alternate method of contacting the help desk. By clicking on the Rescue icon the user is presented with a form which they complete and send to the help desk. Rescue also collects copies of system files (win.ini, config.sys, autoexec.bat, etc..) from the user's computer. These can be used by the help desk in diagnosing the user's problem.

VycorWeb v3.10: The VycorWeb product enables help desk customers to log and check the status of open tickets via a browser on the World Wide Web. Additionally help desk technicians and analysts can access, update, and close tickets remotely via the Web.

The system is integrated with Inference's CasePoint, CBR Express and Crystal Reports.

Metrix, Inc.

Corporate Headquarters:
20975 Swenson Drive
Waukesha, WI 53186

Phone:
414-798-8560

Fax:
414-798-8562
800-543-2130

E-mail:
info@metrix-inc.com

Web:
www.Metrix-inc.com

Products:
OpenUPTIME suite of Service applications
Field Service; Repair Center; Support Desk Techlink for Windows

RDBMS / Environment: Oracle, Sybase, Informix, SQL Server
Operating Systems: HPUX, DOS, SUN OS, Solaris, DGUX, AIX, VMS,
VAX, ULTRIX, SCO UNIX
Hardware Platforms: HP, Sun, Data General, IBM, Digital, Pyramid,
Windows NT, LAN

Product Description:
The OpenUPTIME Support Desk includes an integrated system for tracking
problem reports, requests, and resolutions. Features include a knowledge
database accessed by keyword search for problem resolution. Alert warnings for customer call-back, escalation management, electronic queuing,
and a "linking" capability for one or more requests linked to a problem
report. OpenUPTIME supports Automatic Number Identification
(ANI)/Caller ID to quickly identify your callers.

OpenUPTIME Field Service product is a comprehensive package for call
taking, scheduling, paging, and dispatching service engineers.
OpenUPTIME Field Service incorporates contracts, preventive maintenance, warranties, inventory, and invoicing. With OpenUPTIME you can
automate contract renewals, generate contract renewal letters and perform
parts order entry. OpenUPTIME includes dispatch/schedule board for point
and click re-assignment of calls.

OpenUPTIME Repair Center manages RMA requests, receiving, inventory,
repair, shipping and billing associated with depot repair at medium - to large-
sized organizations, in localized or multi-site environments. The
OpenUPTIME Repair Center also manages warranties and service contracts, shop floor routing, and reporting. Business rules act as guidelines

that determine what processes (billing, shipping, receiving, repair, etc.) occur at which repair sites. Repair Center include high volume RMA issue, and supports bar coded data entry for rapid input of parts and products.

OpenUPTIME Techlink for Windows is the laptop extension of OpenUPTIME. With Techlink, technicians can download a copy of their service calls to the laptop (including place, products installed, configurations, history, etc.), perform the service, print a service report, and upload the call information back to the executive database

Each module takes advantage of full integration for tracking warranties, contracts, call management, service agreements, field and warehouse inventory management, and financial review and cost reporting. OpenUPTIME automates service schedules, troubleshooting, preventive maintenance, engineering change orders (ECOs), and RMA management. OpenUPTIME can address multiple languages & currencies.

The Molloy Group

Corporate Headquarters:
Four Century Drive
Parsippany, NJ 07054
Phone:
201-540-1212
Fax:
201-292-9407
E-mail:
molloy@molloy.com
Web:
www.molloy.com

Products:
TOP Of MIND

All TOP Of MIND products run on Windows 3.1 or higher; are Windows 95 compatible; and the client/server version supports all ODBC-compliant SQL databases.

Products and Prices:
TOP Of MIND Standard Help Desk
for Windows: . . . .$13,200 for five concurrent users (LAN);
. . . .$1,100 per user after 15.

TOP Of MIND Advanced Help Desk
for Windows: . . . .$18,000 for five concurrent users (LAN),
. . . .$1,500 with more than 15 users.

TOP Of MIND Client/Server Help Desk
for Windows/SQL: prices vary depending on number of users.
. . . .$26,500 for five concurrent users
. . . . .decreasing to $2,000 per user for 21 to 75 concurrent,
. . . .$1,000 for over 100.

TOP Of MIND Features and Options
Cognitive Internet(tm) . . . .$225 per concurrent user

Knowledge Kiosks(tm) .$12,000 Server plus up to 5 concurrent users.

Knowledge Cubes   . . . .$500 per concurrent user
(FoxPro version) . . . . $750 per concurrent user
(SQL version) . . . .$180 Annual Subscription Agreement
(includes updates and support)

Support On Site   . . . .$2395 3 concurrent users
. . . .$4995 5 concurrent users
. . . .$25,000 Site license (unlimited use)

Product Description:

Molloy Group markets TOP Of MIND Help Desk for Windows, which uses a proprietary problem diagnosis and resolution technology called the Cognitive Processor to accumulate and distribute the collective knowledge of the help desk staff. This system, in effect, learns from the experience of its users.

TOP Of MIND integrates call logging, tracking, asset management and problem resolution for internal and external help desks and customer service environments. The Cognitive Processor, (patent pending) based on a hybrid of neural networks, fuzzy logic, text-matching and other techniques, builds knowledge "on the fly." The Cognitive Processor maintains performance as the database grows.

Cognitive E-mail(tm) lets the system respond to e-mail messages, even via Internet, by automatically generating call tickets for each case. Calls can be "smart routed" to the proper second-level technicians based on past experience. Links to WinBeep(tm) provide remote dispatch capability. Hypermedia is integrated throughout including word processing, sound, video and picture files. Hypertext area accommodates a decision-tree of multiple layers for extensive on-line documentation. Escalation feature prioritizes open calls and routes e-mail notifications to appropriate parties.

Molloy Group also provides proprietary Knowledge Cubes(tm) — packaged knowledge content products featuring trouble-shooting data on 21 popular hardware and software products and technologies. The knowledge is processed into a data structure that takes advantage of the Cognitive Processor. The content for the Cubes comes from knowledge products marketed by ServiceWare Inc., under a unique co-branding agreement between Molloy Group and ServiceWare.

Along with the knowledge content, Molloy Group has an application called the TOP Of MIND Knowledge Integrator(tm), which manages the importation of the Knowledge Cube into the user's existing TOP Of MIND knowledge base. Through a self-explanatory process, the Integrator directs the loading of the Knowledge Cube data into TOP Of MIND, reconciling the standardized Cube data with the user's file structure, ensuring that records are stored in the appropriate tables regardless of changes the user may have made to the data structure. These Knowledge Cubes are dynamic - they grow with use. A typical Cube contains approximately 2000 "cases" representing problem resolutions, answers to common "how-to" questions and more. Also included are graphics, showing error messages, screen captures, etc., which are relevant to specific problems.

Knowledge Cubes can be used with their Knowledge Kiosks(tm) — applications designed to allow access to the knowledge of the Cognitive Processor via the World Wide Web. Knowledge Kiosk technology lets users retrieve knowledge from Knowledge Cubes using any Web browser on any hardware.

Molloy Group's Top of Mind uses hypermedia (pictures, graphics, video) to document problems and solutions to help users find the right information.

The cases in TOP Of MIND establish connections, (weighted associations) between elements such as the problem, the diagnosis and the resolution. TOP Of MIND integrates knowledge into its cognitive network and associates it with existing knowledge.

Molloy Group also has an agreement with Logical Operations, a subsidiary of Ziff-Davis Publishing, to provide access from within TOP Of MIND Help Desk to Support On Site, technical troubleshooting data published in CD-ROM form.

Support On Site for Applications and Support On Site for Networks contain technical information provided by more than 50 major hardware and software vendors. The CD-ROM is sold by annual subscription, and is updated monthly. A mouse-click launches Support On Site from within TOP Of MIND.

Oasis Technology, Inc.

Corporate Headquarters:
601 Daily Drive, Suite 208
Camarillo, CA 93010

Phone:
805-445-4833

Fax:
805-445-4839

E-mail:
oasis@asyst.net

Products:
The OASIS System for Customer Service; The OASIS System for Help Desk; The OASIS System for RBOCS; The OASIS System for telecommunications; The OASIS System for municipalities; The OASIS System for call tracking; The OASIS System for legal case management The OASIS System for power providers; The OASIS System for Natural Gas providers The OASIS System for Water providers The OASIS System for pharmaceutical manufacturers.

Suggested List Price:
$7,259 - 8 users
$17,520.00 - unlimited users

Compatible with (network): Windows for Workgroups, Windows 95 Network, Windows NT, Novell, Lantastic (all versions), Banyan Vines

Compatible with (platform): Windows 3.x, Windows 95

Product Description:

• Automatic call history • Ability to process and manage associated calls • Ability to process and manage external documents such as word processing document, e-mail, voice mail • Task assignment and follow up • Prioritization of calls • Prioritization of assigned tasks • Management of chronic problem areas • Trend analysis • User customizable screens • Ad-hoc report writer • Mail merge capabilities

ONYX Software Corporation

Corporate Headquarters:
330120 Avenue NE
Bellevue, WA 98005

Phone:
206-451-8060

Fax:
206-451-8277

E-mail:
info@onyxcorp.com

Web:
www.onyxcorp.com

Names of Products:
ONYX Customer Center; ONYX Customer Center - Unplugged; ONYX Web Wizards; ONYX Seeker

Suggested List Price:
$5,000 (10 user server license) + $2,500 per user

Compatible with (network): Microsoft Windows NTTM, NetWareTM

Compatible with (platform): Microsoft Windows NTTM, Microsoft SQL Server for Windows NTTM

Product Description:
Designed exclusively for Microsoft BackOffice, ONYX Customer Center manages all enterprise customer interactions from a single application and interface. It combines functions for sales opportunity management, marketing, customer service, technical support and quality assurance.

Using the system, companies can track all opportunities and support incidents associated with a customer and that customer's company. All employees can at a glance, view this information and thus sell to and serve the customer more effectively and efficiently. ONYX Customer Center is not a repositioned help desk or sales force automation product. ONYX Customer Center was designed from the ground up with the customer as the focus — not the opportunity or the incident. The innovative List Management capability also empowers marketers with the ability to more effectively market to their existing customer base as well as to their prospects.

Specific ONYX Customer Center features include:

TeleSales: Offers features like sales opportunity management, activity management, automated literature fulfillment, telephony integration and flexible reporting and analysis. Lets the entire enterprise (inbound, outbound and field sales) share information

Customer Service: Helps streamlined customer service by integrating and automating the process.

Internet Integration: ONYX Web Wizards lets customers access, browse, retrieve and exchange information via the Internet; Enter call information via an Internet Web form, minimizing resources for support call entry; View customer-support history including the current status of all issues; Access the knowledge database via key word or free text searches using an Internet form; Request delivery of tech notes and other product information via e-mail, fax or mail

Quality Assurance: For integrating engineering with marketing and sales. Customer feedback is recorded and shared instantly to ensure full accountability.

ONYX's Customer Center manages all enterprise-wide customer interactions from a single user interface. The system is just as useful for sales and other departments as for the help desk.

Opis Corporation

Corporate Headquarters:
1101 Walnut St., Suite 350
Des Moines, IA, 50309

Phone:
800-395-0209

Fax:
515-284-5147

E-mail:
opis.com

Products:
SupportExpress SupportExpress Help Link SupportExpress Telephony Link SupportExpress Remote Control Link

Suggested List price:
One user system $2,995
Five user system $6,995
One additional user $1,495
Five additional users $4,995

Compatible with (network): Runs on any IPX, TCP/IP or NetBUEI network—All major network operating systems

Compatible with (platform): Windows 3.1 or higher, Windows 95 and Windows NT

Product Description:
SE automates the tracking and resolution of customer requests, such as those typically received in customer support, customer service, or technical support departments.

The system is a customer request tracking and resolution system suitable for Workgroups and medium sized companies. SE can also manage customer support contracts, RMAs, and customer inventory. SE is fully customizable and integrates with all popular e-mail systems. SE includes a robust NetWare or NT based, industry standard, SQL database engine and graphical report writer. An optional client module allows end users to submit tickets and query the status of their tickets. Other modules integrate SE with telephone switch third party remote control packages. Offers 32 bit application to take advantage of new operating systems and CPUs.

Database server module - Provides secure, scalable data repository. Immune to server crashes.

Customer Request tracking module - Provides a way to track calls, customers history, customer activity.

Calendar and to-do module - A personal information manager for the help desk operator and/or executive information system for the administrator

Customer Request resolution module - standard procedures, standard problems, knowledge base and SpeedSearch. (SpeedSearch uses text searching technology from Verity - the leader in text search tools) Integrate with and use external knowledge bases from ServiceWare.

Other Modules: Customer inventory management module; Customer Support contract management module; Customer RMA management module (to track product returns); administration module; Workflow module (to automate tasks); report writing module.

Links to remote control applications - ReachOut or Proxy Links to telephone systems - Link to any TAPI or TSAPI compliant switch.

Client help module - Reduce discovery time, provides 24 hour customer access to the help desk, turn demand support to scheduled support

Verity text searching technology - the industry standard in full text searching (boolean, phrase, sentence, paragraph, many, accrue, wild card, basic zone, stemming).

This is a completed detailed incident (listing problem and solution) screen, part of Opis' SupportExpress.

Peregrine Systems, Inc.

Corporate Headquarters:
12670 High Bluff Dr.
San Diego, CA 92130

Phone:
800-638-5231
619-481-5000

Fax:
619-481-1751

E-mail:
info@peregrine.com

Products:
ServiceCenter is overall help desk product line. Applications include:;
Problem Management application; Inventory Management application;
Change Management application; Order & Catalog Management applica-
tion: Service request processing; Financial Management application; IR
Expert: Neural net solution rediscovery; IPAS: Interface to OpenView and
NetView/AIX for dynamic status and inventory; SC Automate for SPEC-
TRUM: Interface to SPECTRUM for dynamic status and inventory; SC
Automate for SunNet Manager: Interface to SunNet Manager for dynam-
ic status and inventory; SC Automate for Lotus Notes: Allows Lotus
Notes users to open, update and close problem tickets; NAPA: Interface
to NetView/MVS for dynamic status and inventory and job abend man-
agement; AXCES: Interfaces ServiceCenter with external applications

Suggested List Price:
$9120 for Problem Management and three named users

Compatible with (network): Client/Server product; requires TCP/IP
connection to clients

Compatible with (platform): Servers: MVS, UNIX, Windows NT
Server, NCR, RS/6000, AIX, Sun OS/Solaris, HP-UX.

Clients: Windows 3.1, Windows 95, Windows NT, OS/2, Motif,
Macintosh, Web. Databases: DB2, Oracle, Ingres, Sybase, MS SQL
Server, Peregrine P4

Product Description: ServiceCenter is an integrated suite of applications for automating help desk support services including: call, problem,
change, asset, financial, and service request management across a heterogeneous, enterprise IT environment. The products enable logical automation of service and administration across multiple platforms, including SNA,
various UNIX flavors, and Windows NT, from a single GUI workstation.

ServiceCenter is based on object-oriented technology and employs a three-tier client/server architecture comprising a central data repository, a process modeling engine, and a graphical front-end, each maintained independently and in communication with the other components. This architecture enables administrators to make changes to one element without having to reconfigure the others (for example, changes to the workflow engine do not require manual changes to affected forms on the GUI).

ServiceCenter includes support for all popular SQL databases; an intelligent neural net solution rediscovery tool that enables support staff to perform ad hoc queries; customizable reporting capabilities and interface screens; and integrated security features.

ServiceCenter's distributed agent technology improve both help desk and end-user notification by publishing management data to one or more "subscribing" objects (such as a chart or marquee) at the GUI, providing real-time updating of status information. Distributed clients and servers running on multiple platforms can interoperate.

Five integrated applications are available as part of ServiceCenter: — Call and Problem Management: Provides proactive problem management, including trouble-ticketing, automated response, and audit trail logging. Automatically detects problems, populates trouble-tickets with inventory and asset data, notifies appropriate staff, escalates unresolved problems, and closes resolved tickets. Used in conjunction with Peregrine's interfaces to NetView, OpenView, SunNet Manager, Spectrum, Tivoli TME, CommandPOST and MAXIM, the product collects alerts from these management platforms and assigns problems in real- time. These interfaces also allow automatic discovery of network inventory and loading of the information into ServiceCenter files. Reporting features show downtime for a particular device, total network downtime as a result of a problem, and downtime during business hours. The product enables administrators to submit simple language queries to generate a list of possible resolutions from the database.

Change Management: Provides electronic documentation and monitoring of the change process from proposal through acceptance, scheduling, approval, review, coordination, and cost accounting. Administrators can set up change models containing a pre-defined set of tasks for executing a change; these models can then be adapted to particular change projects. The product automatically calculates risk and identifies risk areas.

Asset Management: Provides a central repository for information on hardware devices, components, and subsystems, as well as software, staff, and remote equipment. Automatically generates topology displays and provides tracing from any network element. Integrates with the other ServiceCenter modules, automatically supplying data to support these processes. An object-based data repository enables definition of hierarchical and peer-to-peer network connectivity, to facilitate problem determination and change planning.

Financial Management: Provides central management of financial informa-tion on hardware, software, and assets. The product provides an efficient, consistent method of handling contracts, supplier invoices and cost allo-cations and preparing budgets and forecasts.

Service Request Management: Automates and tracks the process of ordering products and services, from initial request to purchase/work orders, to delivery and distribution of the products; maintains a catalog of approved products and services for user reference. Enables consolidation of purchase orders to achieve volume discounts from vendors, tracking of existing inventory to eliminate unnecessary orders, and tracking of replaced equipment to enable resale or trade.

In addition to ServiceCenter core functionality, Peregrine provides a suite of tools to enhance and extend the product's capabilities. They include:

IR Expert is an expert system using neural network technology to provide access to past problem resolutions from internally developed or externally supplied knowledge bases. The product allows intuitive, common-lan-guage-based queries of the knowledge base to speed problem resolution. It is a self- learning tool, and requires no maintenance to operate.

AXCES is a generic API which provides a seamless data interface between ServiceCenter and external applications, to ease integration. Events trig-gered by external applications are transferred by AXCES to ServiceCenter, which automatically opens, updates, and closes problem tickets and updates inventory records based on the event. The two-way interface also intercepts ServiceCenter events and provides notification via pager, e-mail, or other management applications.

SC Automate is a set of tools for integration of functions with popular host and open management platforms and applications, such as Cabletron Spectrum, SunNet Manager, and Lotus Notes.

PLATINUM Technology, Inc.,
Answer Systems Laboratory

Platinum headquarters:
1815 South Meyers Road
Oakbrook Terrace, IL. 60181

Phone:
800-442-6861
708-620-5000

Fax:
708-691-0710

E-mail:
info40platinum.com

Web:
www.platinum.com

Answer Systems headquarters:
2 N. Second St., Suite 1000
San Jose, CA 95113

Phone:
800-677-2679
408-280-5110

Fax:
408-280-1004

E-mail:
info40answer.com

Web:
www.answer.com

Products and Prices:

Apriori GT (Problem resolution & call tracking): 5 users, ...37,500
10 users 70,000
Unlimited users, ...320,000
Apriori LS: 10 users . . . .
(LS is an entry-level system for small help desks only.)
Apriori PLUS (Problem Resolution only): 10 users, ... 40,000
Unlimited users, ...280,000
TextTracer (option): five users $14,300
10 users $28,000
Apriori Hands-Free (option): $15,000

Servers Supported: HP 9000, IBM RS/6000, Sun OS/Solaris

Databases Supported: Sybase, Oracle, Native Relational Database

User Interfaces Supported: PC with Windows 3.x, Windows NT Windows 95, X Window System (Motif and OpenLook), Character-based interfaces, Apriori Hands-Free Support (optional World Wide Web/e-mail interface)

Product Description:

Apriori GT: The complete help desk automation system, from call management to problem resolution, with workflow management, and reporting.

Apriori PLUS: The problem resolution system for help desks that already have a call management system in place. It includes all the workflow management, and reporting features of GT.

Apriori GT and PLUS owners can add options ranging from Hands-Free Support, which enables users to search the Apriori knowledge base via e-mail or the World Wide Web, to external text-retrieval engines, prepackaged knowledge bases, defect tracking interfaces, and integration with Hewlett-Packard Company's OpenView IT/Operations. Products use three-tiered client/server architecture and industry-standard TCP/IP and WinSock network protocols.

Primus Communications Corporation

Corporate Headquarters:
1601 Fifth Avenue, Suite 1900
Seattle, WA 98101

Phone:
206-292-1000

Fax:
206-292-1825

Products:
SolutionBuilder 2.0, SolutionPublisher 1.0

Price:
SolutionBuilder: $3,750 per seat

Compatible with (networks): Sun Solaris, Microsoft NT

Compatible with (platform): Sun Solaris, Microsoft NT, Microsoft Windows 95, X-servers

Product Description:
SolutionBuilder is a support automation system that enables Solution-Centered Support, a strategy developed by the Customer Support Consortium. The system lets support analysts author solutions in real-time.

SolutionBuilder eliminates the need for knowledge engineering efforts required with traditional case-based reasoning systems. SolutionBuilder can integrate with existing call management and text retrieval systems to provide a complete support center solution.

SOLUTIONPUBLISHER: For companies using Primus SolutionBuilder, SolutionPublisher provides clients and partners with direct access to solutions over the Internet and corporate intranets. SolutionPublisher is a Web client server that can be accessed 24 hours a day, 7 days a week, by anyone with Netscape Navigator or another World Wide Web browser. Companies move specified solutions from SolutionBuilder to the SolutionPublisher server, to ensure corporate security and safety.

ProAmerica Systems

Corporate Headquarters:
959 E. Collins
Richardson, TX 75081

Phone:
800-888-9600
214-680-9600

Fax:
214-680-6134

E-mail:
bruceb@proam.com

Web
www.proam.com

Products:
Service Call Management software Web module RMA module Defect Management module Toolkit

List price:
$1895 per concurrent user

Compatible with: UNIX, Windows NT, Novell, OS/2 database servers; Windows, NT and Windows 95 client desktops; Informix, Oracle, DB/2 6000, Sybase, Microsoft SQL database engines.

Product Description:
Service Call Management (SCM) helps support reps track, analyze, manage and respond to support requests and measure customer satisfaction. Users can execute commonly used queries or build new ones. The system can monitor call status and automatically escalate calls, track calls by many types of criteria, verify product registration and check contract agreements.

The Defect Management module, used by software developers, works with SCM to support and manage the software product lifecycle.

The Web module works with SCM to let customers access the system the same way support reps do. They can log a support call, review the status of that call, access the knowledge base to find their own resolution or e-mail a support rep.

Professional Help Desk

Corporate Headquarters:
800 Summer Street
Stamford, CT 06901

Phone:
203-356-7900

Fax:
203-356-7900

E-mail:
info@prohelpdesk.com

Web:
www.prohelpdesk.com

Products:
PHD-Professional Help Desk; Total Asset Management (TAM); ClientView
Internet Access

Platform Compatibility: any platform running an ODBC supported
database, including UNIX, OS/2, VAX and mainframes; Operating Systems:
Windows '95, 3.1, NT, Windows for Workgroups & OS/2 for Windows.

Network Compatibility: Novell Netware, Windows NT Server,
Windows for Workgroups, Pathworks, Banyon Vines, Lantastic.

Product Description:
PHD-Professional Help Desk is an integrated help desk system with call
tracking, call management, problem management and problem resolution
features. PHD's Natural Intelligence suite consists of Rapid Resolution,
EBR, Natural Language and PHD Expert.

The system uses a proprietary fuzzy logic based system called EBR
(Experience Based Reasoning) that learns from experience. The system can
also use a statistical methods match, case based reasoning and Natural
Language (their patented technology) understanding.

Total Asset Management allows help desks to manage enterprise assets
and provides inventory support.

Prolin

Corporate Headquarters:
Two Stamford Plaza, 281 Tresser Blvd.
Stamford, CT 06901

Phone:
203-406-1236

Fax:
203-406-1239

Web:
www.prolin.com

Products & Pricing:
Configuration Manager . . . .$7,500 server
. . . .$2,000 per user (1-25)
Helpdesk Manager; Problem Manager; Change Manager; Service Level
Manager; Software Control & Distribution; Notifier/Auditor; World
Wide Web Interface

Server prices the same for each module . . . .$1,700 per user (26-50)
. . . .$1,400 per user (51-100)
. . . .$1,100 per user (100+)

Compatibility: Oracle7 Server running on UNIX, Windows NT or VMS.

Product Description:
Configuration Manager - manages relevant assets and the relationships
between them

Problem Manager - supports the registration dispatching and tracking of
problems. Retrieves information about configuration items, service calls
and changes. Helps determine the underlying cause of failure.

Helpdesk Manager - allows the registrations, dispatch and tracking of service calls.

Change Manager - supports requests for changes, impact assessment,
authorization, building, implementing and evaluation of changes within the
IT infrastructure.

Quintus Corporation

Corporate Headquarters:
47212 Mission Falls Ct.
Fremont, CA 94539

Phone:
510-624-2800
800-337-8941

Fax:
510-770-1377

E-mail:
sales@quintus.com

Products:
CustomerQ; WebQ; Design Tool

Prices:

CustomerQ . . . .$4,000 per user
. . . .$20,000 per server
WebQ . . . .$20,000 per server

Compatible Network: TCP/IP

Compatible Platform: Servers supported: HP 9000, IBM RS/6000, Sun SPARCstation OS: HP/UX, AIX, Sun Solaris Clients Supported: PC—Windows 95, Windows 3.1, Windows NT, Windows for Workgroups, Any Web Browser Databases: Informix, Oracle, Sybase

Product Description:
Users can retrieve solutions stored anywhere in the enterprise. CustomerQ's problem resolution is an intuitive search mechanism that examines both database text records and documents, retrieving solutions ranked by relevancy. CustomerQ searches solution collection by default. If the answer cannot be found there, agents can broaden their search for documents stored anywhere in the enterprise. Full-text capabilities are provided by CustomerQ's integration of Fulcrum's SearchServer engine. Solution documents may be viewed directly within CustomerQ while retaining their native formatting and appearance.

Radish Communications Systems, Inc.

Corporate Headquarters:
5744 Central Avenue
Boulder, Colorado 80301

Phone:
303-443-2237

Fax:
303-443-1659

E-mail:
VoiceView@radish.com

Products & Prices:
VoiceView Help Desk: . . . . .$55,500 annual subscription for 10 seats
VoiceView Service Link Server $166,200
(dependent on exact configuration)
InsideLine: . . . .$149.95 per seat
(typically sells for $99 each in system sales).

Compatibility: VoiceView technology is fully compatible with call center ACD, PBX, and key line systems, as well as both analog and digital lines.

Product Description:
Radish offers a combination of products to provide both call deflection and call efficiency in customer support call centers. The first goal is to leverage investments in existing call center equipment. Next the goal is to handle as many calls as possible with capital equipment versus live agents. Finally, when live agents do take a call, the goal is for the system to take the "blindfold off of the agent" using integrated voice and data. This maximizes their efficiency with every caller.

INSIDELINE InsideLine, is a modem adapter for digital phones that solves the difficulty of economically making modem connections in customer support call centers. Separate analog lines for each agent is a costly   alternative and one that does not leverage the existing digital phone lines. The challenge has been to leverage the substantial investment in   digital lines and telephones associated with PBXs and ACDs. Each   InsideLine modem adapter used on an existing PBX/ACD line, avoids pulling a separate analog line.

The device makes it possible for all agents to use a VoiceView modem with the company's existing proprietary digital phones, while preserving the utility of all of the value-added features of that phone and associated switching equipment.

VOICEVIEW SERVICE LINK SERVER The VoiceView Service Link server is an automated call deflection system. This server is a Visual Voice Response system. It uses integrated data and voice to remove the traditional limita- tions of conventional VRUs. Rather than long lists of voice prompts with deeply nested menus the system provides easy to navigate screens supplemented by voice. This system answers calls into a customer support call center and interacts with the calling clients to answer their questions and to solve their operational issues directly. For example, top ten questions are easily presented and answered. Driver updates can be done automatically during a single inbound call without the costs of preparing or mailing diskettes. Diagnostics can be run and results analyzed.

Intelligent routing is done if the VoiceView Service Link Server is unable to answer the callers question. When the calling client is transferred to a VoiceView Help Desk equipped support technician the call can also include the diagnostic history performed by the server up to the point of transfer.

VOICEVIEW HELP DESK VoiceView Help Desk is a Windows application optimized for customer service agents. It is designed to significantly shorten technical support calls. It does this by providing integrated voice and data over a single phone line. This means that the interaction between VoiceView-enabled call center agents and VoiceView-capable clients can incorporate specific technical support features only available with a voice and data link. Now a service technician can make remote requests of the caller's computer during the incoming voice call. For example, the technician can request a file, ask to have a file saved to a particular place, request an application to run or get a screen capture.

Once security is established with the caller then VoiceView Help Desk documents each step of the session. It does so in an on-screen log file on the service representatives' system. The system also presents automatically generated information such as caller identity, product serial number, problem description and system configuration. It supplements this log with the files sent and received, etc. Rollback recovery to all caller supplied files is supported automatically. Linking to other call center databases is supported.

Additionally, when calls are escalated within the call center the log files may be sent to other technicians via a simple phone transfer. This uses the existing call center switching equipment without any change. VoiceView Help Desk also allows call center staff to develop and run custom macros that automate sequences of actions. Extensive remote control of client machines and remote screen capture is also supported.

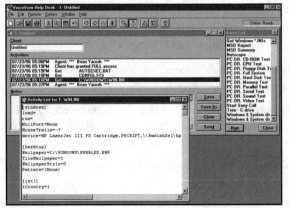

This screen comes from Radish Communication's VoiceView, a product that lets users remotely transfer data from PC to PC over a single phone line.

Remedy Corporation

Corporate Headquarters:
1505 Salado Drive
Mountain View, CA 94043

Phone:
415-903-5200

Fax:
415-903-9001

E-mail:
info@remedy.com

Web:
www.remedy.com

Products:
Action Request System (AR System); Flashboards; ARWeb; Change Management Solution; Asset Management Solution; Multi-Processor Server Option (for NT); Distributed Server Option

Suggested list prices:
ARSystem $6,500
Flashboards $5,000
ARWeb $12,000
Change Management . . . .$5,000/5 fixed user licenses
Asset Management . . . .$5,000/5 fixed user licenses
Multi-Processor Server Option $3,000
Distributed Server Option $5,000

Network Compatibility: ONC RPCs on TCP/IP (runs on LANs {ETH-ERNET, token ring, etc...} and on WANs {dial-up and permanent circuit})

Network Management Compatibility: AT&T OneVision, Boole&Babbage CommandPOST, Bull Integrated System Manager, Cabletron Spectrum, HP OpenView, NetView for AIX, Solstice Site Manager, Bay Networks, Tivoli Management Environment

Client platform compatibility: Windows 3.X, Windows 95, Windows NT, Macintosh, Web Browsers, ASCII, E-mail.

Server platform compatibility: NCR System 3000, Hewlett Packard 9000, Windows NT, IBM RS/6000, Motorola mc88100, Motorola PowerStack, Silicon Graphics, Sun SPARC/Solaris and Sun SPARC/SunOS.

Database compatibility: Informix, Ingres, Microsoft SQL Server, Oracle, Sybase, also supports flatfile!

Product Description:
Remedy's AR System lets customers automate support processes. Using a three-tier client/server architecture, Remedy allows companies to lever-

age existing technology investments by providing an extensive array of integrations.

Remedy customers can open the box and get started in as little as 30 minutes, says the company. However, if the user desires a more custom-fit solution, (i.e. if they want to integrate a home grown application they have in house), Remedy offers its APIs at no cost with the system, allowing complete customization if desired.

Remedy's newest version of its ARWeb software, released July 1996, adds enhanced security and remote access capabilities, allowing the product to automate Internet support.

Flashboards, allows companies to monitor the performance of the AR System. Using a series of meters and multi-line charts, Flashboards tracks data stored in the AR System, providing alarms and notifications if warning levels are reached.

For example, if a user sets up Flashboards to track critical calls, and wants to ensure that all such calls are closed within 30 minutes, then the user can set an alarm for any time prior to the 30 minute window. Flashboards will continually poll the AR System, and if any data begins approaching the warning level, the application will set off a notification (beeper, e-mail or notifier tool) ensuring that the danger level is noted so it can be resolved.

Repository Technologies, Inc.

Corporate Headquarters:
6825 Hobson Valley
Drive Woodridge, IL 60517

Phone:
800-776-2176
708-515-0780

Fax:
708-515-0788

Web:
custfirst.com

Products:
CustomerFirst; ControlFirst; OrderFirst

Suggested List Price:
5 - user base system CustomerFirst $9,995
5 - user base system ControlFirst $4,995

Compatible with (network): Netware, NT Server, Windows for Work Groups, Banyon Vines, LAN Manager

Compatible with (platform): Windows 3.x, Windows/NT, WIN95, WINOS/2

Product Description:
CustomerFirst: Call tracking & problem management system with automated rule-based escalation; workflow management; task allocation; notification when maintenance has expired; tracking of all fixes and source code changes to each version; SQL database.

ControlFirst: Integrated problem tracking and management system with workflow management; interface to INTERSOLV's PVCS source control system; SQL database.

OrderFirst: System for tracking sales quotes, orders, order fulfillment, shipping confirmation, invoicing, processing payments or credit card processing order adjustments, processing warranty registration cards and maintenance renewal orders.

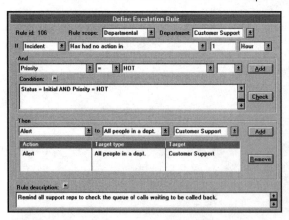

In this screen, part of CustomerFirst from Repository Technologies, a rule is defined that will alert the entire customer support department if no support rep has called the customer bask after one hour.

Scopus Technology, Inc.

Corporate Headquarters:
1900 Powell Street
Emeryville, CA 94608

Phone:
510-597-5800

Fax:
510-597-5994

E-mail:
info@scopus.com

Web:
www.scopus.com

Products:
SupportTEAM; SupportTEAM Plus; SalesTEAM; QualityTEAM; Scopus
Voyager/Lift Off; ServiceTEAM; WebTEAM; WorldTEAM; NetTEAM;
TeleTEAM; ScopusWorks; CommLink

Suggested List Price:
$3000 - $5000/concurrent user (includes both client and server pricing)

Compatible with (network): TCP/IP

Compatible with (platform): HP, SUN, IBM, SGI, NT

Product Description:
While the term "help desk" is often used to describe any corporate support
system used to resolve customer problems — whether from internal "customers" (i.e., employees) or a firm's actual paying customers — Scopus
applications are designed to serve the very different purposes, functions and
requirements of an internal help desk versus an external call center.

From a business perspective, the internal help desk is aimed at maximizing
efficiency within the corporation and better utilizing its information technology assets. In an era when the cost and complexity of information technology are growing exponentially even as IS budgets are decreasing, Scopus'
help desk application focuses on improving call management, problem resolution, resource management (e.g., tracking internal network and machine
configurations), change management (including the administering of IS
upgrades, scheduled outages and network reconfigurations), Service
Delivery Management (SLAs) and third party dispatch, and workflow
automation. Calls to internal help desks tend to be highly-technical in
nature. The external call center, on the other hand, is aimed at maximizing
the company's effectiveness in the marketplace.

By better serving customer needs, call centers provide strategic advantage
to the enterprise by enhancing customer loyalty and by turning support

Here's an example of how Scopus' WebTeam can be used to let user' of Scopus' products let their own customers submit a problem ticket and/or search their knowledge base over the Internet to find a solution on their own.

inquiries into sales opportunities. For example, when a bank's customer calls to transfer funds from a savings to checking account and is offered the bank's overdraft protection service. Scopus' call center application provides for automated customer call management, case handling and problem resolution, CBR (Case Based Reasoning) searching, product quality and defect handling, RMA and field service dispatch, and the integration of customer support activity with other departments within the enterprise, such as sales and marketing.

Call centers typically handle a much higher volume of calls than do help desks, and these tend to be more transactional than technical in nature. For both help desk and call center operations, Scopus applications offer tools for skills-based routing of incoming calls, CTI, automated database replication throughout the enterprise (including the firm's outsourcing partners), and the integration of multiple technology platforms, such as the Internet and World Wide Web.

ServiceWare Inc.

Corporate Headquarters:
333 Allegheny Ave.
Oakmont, PA 15139

Phone:
800-KPAKS-4-U

Fax:
412-826-0577

Web:
www.serviceware.com

E-mail:
info@serviceware.com

Products:
Knowledge-Pak Desktop Suite(tm)

Suggested list Price: The price of the Suite is $1,500/user from ServiceWare. However, 15 leading help desk vendors are now bundling the Knowledge-Pak Desktop Suite as a standard component in their customer support solution. In those instances, each vendor determines the price for the Desktop Suite.

Compatible with (network): The Knowledge-Pak Desktop Suite is compatible with Novell NetWare and Microsoft Windows NT Advanced Server.

Compatible with (platform): The Knowledge-Pak Desktop Suite is currently included as a standard feature within the following help desk vendors' products: Advantage kbs (IQSupport); Allen Systems Group, Inc. (ASG-IMPACT); Applix, Inc. (Target Helpdesk); Bendata, Inc. (HEAT/First Level Support); Clarify, Inc. (ClearHelpdesk); Datawatch Corporation (Q-Support); Inference Corporation (CasePoint); Magic Solutions, Inc. (SupportMagic); McAfee Associates (Vycor Enterprise); Platinum Technology, Inc. (Apriori); Quintus Corporation (CustomerQ); Software Artistry, Inc. (Expert Advisor); The Molloy Group (TOP OF MIND); Utopia Technology Partners (Utopia Help Desk) ; Vantive Corporation (Vantive Help Desk)

The Knowledge-Pak Desktop Suite is also compatible with the following environments listed, but no distribution relationship exists.

Emerald Intelligence (Resolve); Folio Corporation (VIEWS); Primus Communications Corporation (SolutionBuilder); Remedy Corporation (Action Request System)

Product Description:
The Knowledge-Pak Desktop Suite is a comprehensive collection of knowledge content for today's most popular desktop products, designed specifi-

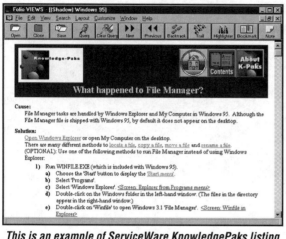

This is an example of ServiceWare KnowledgePaks listing a common problem in Windows 95 and it's solution.

cally for first-level help desk analysts and end users. It contains complete descriptions, causes, and solutions to thousands of desktop-related problems.

Delivered quarterly on a CD-ROM, the Knowledge-Pak Desktop Suite contains knowledge content for widely-used desktop products such as: Microsoft Windows 95; Microsoft Windows 3.1; Microsoft Office Professional 95 Suite; Microsoft Integrated Office Suite; Lotus 1-2-3 Lotus cc:Mail; Lotus Notes; Novell NetWare; IBM OS/2 IBM ThinkPad; Corel WordPerfect PC System Diagnostics; World Wide Web Browsers

The Knowledge-Pak Desktop Suite is comprised of ServiceWare's Knowledge-Paks complete ready-to-use knowledge bases for the help desk which are used by help desk analysts, LAN administrators and end users to solve problems.

Knowledge-Paks contain: - Detailed, step-by-step problem-solving strategies and sample solutions. - Comprehensive lists of error messages and how to recover from them. - Tips on how to optimize performance. - Hundreds of examples, illustrations and annotated screen shots.

Knowledge-Paks are written specifically for first-level support analysts.

Silvon Software, Inc.

Corporate Headquarters:
900 Oakmont Lane, Suite 400
Westmont, IL 60559

Phone:
800-874-5866
708-655-3313

Fax:
708-655-3377

E-mail:
helpinfo@silvon.com

Products:
Helpline (Incident management system) Insite (Asset management system)

Suggested List price:
Helpline: $1,800 per concurrent support representative $300 per concurrent end user Insite: $1.00 per asset up to a maximum of $50,000 for an unlimited license

Compatible with (network): TCP/IP, Digital Pathworks

Compatible with (platform): Servers: Sun Solaris, Sun OS, HP UX, Digital Unix, Digital OpenVMS, Windows NT,

Clients: Windows, Windows 95, Windows NT, Macintosh, HTML, Sun Solaris (Character Cell & Motif), Sun OS (Character Cell & Motif), HP UX (Character Cell & Motif), Digital Unix (Character Cell & Motif), Digital OpenVMS (Character Cell & Motif)

Product Description: Helpline provides the means to manage the workflow of a customer support operation or internal help desk by providing functionality to enter, research, escalate, and resolve incidents within either a centralized or distributed environment. Helpline also provides end-users (customers) with the ability to enter, research, resolve and obtain status without the need to interact directly with support center personnel.

Key features of Helpline include:
Reference Incidents: There may be several incidents which occur repeatedly and therefore have had standard procedures established for their resolution. Reference incidents allow for the inclusion of these standard procedures within the HELPLINE incident database.

Title Search: When a new incident is reported, HELPLINE provides for a rapid search to be performed across the reference incidents for similarities based on problem title. If a match is found, the contents of the previous incident's description, title and resolution may be copied into the new incident occurrence allowing for rapid closure of the new situation. If the contents of the reference incident are used, the new incident is linked as a reoccurrence of the reference incident for easy access and statistical analysis.

Master-subordinate incidents: There are times when a incident is reported that needs to have a team of individuals address the situation. Master-subordinate incidents allow for the separation of the different aspects of an incident and the assignment of each part to the responsible individual. All of the subordinate problems are linked to the master and share information including each subordinate incident's status. HELPLINE may be configured such that the master incident cannot be closed until all of the subordinates have been resolved.

Research Agent: If an incident is saved in the database without being resolved, one of the initial tasks that the assigned support representative generally performs is to search the incident database for similar issues. The HELPLINE Research Agent uses an artificial intelligence algorithm to automatically perform this search in the background. When the incident is opened by the support representative to begin the resolution process, a weighted list of possible matches is provided thereby shortening the amount of time that must be spent on historic research.

GUI: HELPLINE's graphical user interface provides point-and-click ease of use to perform all support and end user functions. The GUI is consistent across platforms. Functions can be performed using either the keyboard or the mouse. An extensive on-line help system provides context-sensitive help, conceptual information, and how-to procedures.

End user problem entry: HELPLINE provides an interface for users to submit problems, get status updates, and search the database for past solutions. It automatically notifies users when problems are assigned and closed. You can define user authorizations to limit access to functions of the end user interface. This capability may be accessed through any supported platform including the Internet.

Support functions : Staff members use support functions of HELPLINE to enter, assign, update, close, delete, reassign, reactivate, transfer or return problems, and make diary entries. Staff members can set their own preferences, such as field defaults and the sort order for problem lists. You can define support authorizations to limit access to these functions.

Support rep workload : Before assigning a problem, you can quickly check the number of problems assigned to each staff member. This enables you to assess their workloads and distribute problems to support staff evenly.

Problem type certification: You can specify which problem types each member of the support staff is authorized to work on.

Elapsed and actual time: HELPLINE time-stamps each problem as it is logged, assigned, and closed, so you can easily determine the elapsed times between these events. You can also activate the actual time fea-

Silvon Software's Helpline lets you link incidents to known errors so you can track the number of times a particular incident occurs.

ture, which prompts support staff for the amount of time worked on problems whenever they are updated or closed. This allows you to track the amount of time staff members spend working on problems.

Problem history: Once you activate the problem history feature, HELPLINE adds a note to the problem data whenever key events occur or specified fields are modified. This audit trail tracks who was responsible for making changes to the problem record and when it was done.

Pre-allocated reference numbers : As an option, HELPLINE can pre-allocate reference numbers to problems before they are logged. This enables the help desk administrator to provide the caller with a reference number at the start of the call.

Print notification: HELPLINE allows you to notify support staff of new or assigned problems by printer, as well as by electronic mail or broadcast messages.

Interface with INSITE: You can use HELPLINE with INSITE, the asset management system for tracking hardware and software or other assets. For example, when a problem is logged regarding equipment failure, you can query the INSITE database for full information about that piece of equipment. Data from the INSITE database can be automatically read into the HELPLINE problem record to reduce data entry. And you can use Report Painter to generate reports that combine data from the INSITE and HELPLINE databases.

Integration with network managers: Sites using a network manager such as NetView, OpenView, Tivoli, or DECmcc can integrate it with HELPLINE. When the network manager raises an alarm, a problem can automatically be logged in the HELPLINE database. When the problem is solved, HELPLINE sends an "all clear" message to the network manager for that event. If you use the HELPLINE escalation monitor, you can define escalation policies that log an event with the network manager when a problem is escalated. Such communication between HELPLINE and your network manager ensures that system problems are addressed promptly before they cripple operations.

Software Artistry, Inc.

Corporate Headquarters:
9449 Priority Way W. Dr.
Indianapolis, IN 46240

Phone:
317-843-1663
800-795-1993

Fax:
317-574-5867

E-mail:
info@softart.com

Web:
www.softart.com

Products & Prices:
SA-EXPERTISE suite:
SA-Expert Advisor...$15,000/server
.................................$3,500/user; site and enterprise licenses available
SA-Expert Evolution...$20,000/server
....$3,500/user
SA-Expert Foundation Manager...$10,000/server
....$3,500/user
SA-ExpertView $15,000/server
Expert Web ....$15,000/web server for Problem Module;
....$20,000/web server for Diagnostics Module
....$30,000/web server for both modules together
Expert Mail Agent $20,000/server
Expert Access ....$10,000/100 users
Expert Quality $5,000/server
....$1,000/user.

Product Description:
The SA-EXPERTISE suite of software applications links an organization's support center with network management, asset and change management, and end-user empowerment tools.

SA-Expert Advisor incorporates call and problem management with multiple problem resolution technologies. Expert Advisor runs on Windows and OS/2 clients and Unix, Windows, and OS/2 servers. It supports the most common relational databases including Oracle, Sybase, Informix, Microsoft, DB2/2, and DB2/6000.

SA-Expert Evolution and SA-Expert Foundation Manager offer comprehensive change and asset management capabilities, with Expert Evolution managing any changes in the business environment and EFM providing "cradle-to-grave" management of corporate assets through their life cycle.

SA-ExpertView bridges the gap in information exchange between the support center and network management platforms including HP OpenView, IBM NetView, Tivoli TME, Cabletron Spectrum, Boole & Babbage Command Post, and CA-Unicenter.

SA-Expert Web, SA-Expert Mail Agent and SA-Expert Access provide problem resolution directly for end users and customers without analyst intervention through the respective mechanisms of the World Wide Web, e-mail systems, and distributed knowledge.

Expert Web supports any computer with an Internet browser. Expert Mail Agent is compatible with Lotus cc:Mail and Notes Mail, Microsoft Mail, and Novell GroupWise. Expert Access runs on any Windows workstation.

SA-Expert Quality manages the life cycle of customer input in the form of defect reports and enhancement requests.

Expert Advisor is Software Artistry's flagship product which offers call management capabilities and powerful diagnostic technologies. These techniques include Adaptive Learning, Hypermedia Decision Trees, Case-Based Reasoning, Hot News, Common Problems, Quick Solution, and Error Messages. No other support solution offers so many different ways to determine the solution to any given problem.

Other features include:

- cross-platform operability for clients running Windows 3.x, 95, NT, or Workgroups, as well as OS/2.

- server support for Windows NT, OS/2, HP-UX, and Sun Solaris.

- enterprise-wide problem routing and escalation.

- scalable client/server architecture.

- support for major RDBMSs including Oracle, Sybase, Informix, Microsoft, DB2/2, and DB2/6000.

Expert Advisor offers complete enterprise integration through other applications in the EXPERTISE suite in addition to interfaces with legacy mainframe systems and applications and CTI applications and telephony systems.

SolutionDesk

Corporate Headquarters:
1486 St. Paul Ave.
Gurnee, IL 60031

Phone:
847-662-6288

Fax:
847-336-7288

E-mail:
72630,2370@compuserve.com

Product:
SolutionDesk

Suggested List Price:
$499 per user

Compatible with (network): Windows, indows NT, Novell

Compatible with (platform): Windows 3.xx

Product Description:
Internal/external software designed for small to mid-sized businesses.
Main modules include: Incident Inquiry, Literature Request and
Inventory/Asset Management.

Tally Systems Corp.

Corporate Headquarters:
P.O. Box 70
Hanover, NH 03755

Phone:
603-643-1300
800-262-3877

Fax:
603-643-9366

E-mail:
product.info@tallysys.com

Products:
Cenergy NetCensus CentaMeter

Suggested list price:

Cenergy ....$28 per PC for 1,000 PCs
NetCensus..$12 per PC for 1,000 PCs
CentaMeter ....$12 per PC for 1,000 PCs

Compatible with (network): NOS independent; any NOS supporting file sharing and record locking

Compatible with (platform): Windows 3.x, Windows 95, Windows NT, DOS

Product Description:
NetCensus, is an automatic hardware and software inventory product, that links to most major help desk products. NetCensus recognizes hundreds of brand-name PCs, add-ons, and peripherals and identifies software programs by brand name, version number, embedded serial number, and foreign language edition. Cenergy and NetCensus gives help desk managers the complete picture: what you have and where it is, an extensive product recognition library, and customized reporting capabilities.

Teubner & Associates

Corporate Headquarters:
623 S. Main St.
Stillwater, OK 74074

Phone:
800-529-4377
405-624-8000

Fax:
405-624-3010

E-mail:
esp.sales@teubner.com

Name of Product:
ESP - The Expert Support Program

Suggested List Price:
$3,695 (3-user version)

Compatible with (Network): Any Netbios compatible network

Compatible with (Platform): Windows, Windows 95, Windows NT, Macintosh

Product Description:

ESP is a technical support automation application that allows multiple support agents to log and track calls, record and retrieve problem resolutions, and produce reports out of a centralized database. A primary interface, called the Quick Call Screen, allows the support agents to execute all primary tasks from a single screen.

Fuzzy Logic Search technology is employed to manage the storage and retrieval of solutions. Interfaces to both e-mail and fax enable the support agent to deliver solutions and dispatch problems to others who do not have access to ESP. A feature called the ESP Dashboard allows a supervisor to monitor the status of their support operation in real-time through a graphical interface. An interface to the World Wide Web allows support clients with Web browsers to search for solutions in the ESP Knowledgebase.

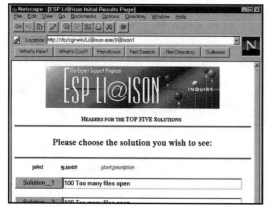

Teubner & Associates' ESP Li@ision lets web browsers access the ESP database.

STEPS: Tools Software,
A Division of Cauchi Dennison & Associates Inc.

Corporate Headquarters:
367 Spadina Road,
Toronto, ON
Canada M5P 2V7
Phone:
416-481-5047
Fax:
416-485-6566
E-mail:
stepstls@msn.com

Products:

STEPS: Tools Service Centerv1.2 STEPS: Tools CMS Plannerv1.2
HelpDesk/Resource Tracking Multi-project/Change Management &
Migration Planning

Suggested List Price:
STEPS: Tools Service Center Lite HelpDesk for single exclusive user
managing no more than 50 nodes ....$595.00.
Service Center Lite Upgrade: Upgrades single-user Lite to manage 100
nodes $360.00

STEPS: Tools Service Center or Tools CMS Planner
Single user $1320.00
5-user site license $5280.00
10-user $10560.00
20-user...$21120.00

Larger site licenses priced individually on a case-by-case basis. Databases
are .mdb files. SQL-Server/Oracle back-end database options are priced
separately and in addition to the above client software

Compatible with (platform): Windows 3.11, Windows95, NT or as a
Windows application running under OS/2 or UNIX. (NTSL tested)

Product Description:
STEPS: Tools is a network enterprise management system. Depending on
the need, users may employ it as an overall and comprehensive network
system or to simply deal with a specific set of tasks at hand, such as hotline support or help desk activities.

The CMS Planner ties related network activities with a data warehouse. It
includes the STEPS publication, customizable project management templates, an integrated on-line system, multi-project management and audit

capabilities. The product automatically creates new project template sets, automatically re-links templates, allows user to add templates, build and manage all project activities in the same system (network specific or not).

In addition to service desk commands, you also have access to several additional items like: - An asset movement tracking system if a device needs to be tracked when replaced or moved for repair; - An alert screen where all assigned, due and late priority items will appear for automatic review (either by a selected individual resource or a global search) - All time and materials billings on closed tickets will post to budgets for relevant accounts (these are handled as transactions and so are readily accessed by corporate accounting systems for other uses) - Reporting capabilities exist for all service center data screens - As an add-on option to the full client-server systems, the system can be set so that all node users are able to send an alert to open a trouble ticket. Such tickets are assigned to the hotline or service desk and alerts are issued so that these tickets can be picked up and assigned.

*STEPS is an acronym for the Structured Transition and Engineering Planning System developed by Cauchi Dennison and published by John Wiley & Sons Ltd.

Vantive

Corporate Headquarters:
2455 Augustine Drive, Suite 101
Santa Clara, CA 95054

Phone:
800-582-6848
408-982-5700

Fax:
408-982-5710

Web:
www.vantive.com

Products:
Vantive Sales; Vantive Support; Vantive Hel Deskp; Vantive Quality;
Vantive Field Sales

Pricing:
Server starts at ....$10,000 per application client
....$2,500 per concurrent user

Compatible with (network): any TC/IP network

Compatible with (platform): Client: NT, Windows, Mac, Motif, Web browser Server: NT, HP, NCR, SUN, IBM, Data General

Product Description:
Vantive Support automates call tracking, problem resolution, and service and support management. The system tracks each case and maintains a complete audit trail. It uses a set of interrelated objects to create and maintain information about customers, contacts, products and agreements. A second set of problem, bug, enhancement and resolution objects are used to track product assistance and enhancement requests.

Vantive HelpDesk tracks and handles employee problems, issues, complaints, suggestions and requests for assistance with technology, resources and the physical plant. Users can submit trouble tickets electronically or by calling the help desk. The system uses a set of interrelated objects to create and maintain information about human resources, escalations, product types and assets. Another set of objects are used to track problems, trouble tickets and resolutions. The system integrates with network management systems and can track and change the status of computer systems automatically if there is a problem with the network.

Here's a problem and it's solution, found using Vantive Support from The Vantive Corp.

Verity

Corporate Headquarters:
1550 Plymouth Street
Mountain View, CA 94043
Phone:
415-960-7600
Fax:
415-960-7698
E-mail:
info@verity.com
Web:
www.verity.com

Products:
Topic family of software tools and applications.

Product Description:
Verity's Topic products are designed for searching, retrieving and filtering in
formation across the Internet, enterprise and CD-ROM. Users can conduct
searches across multiple sources and multiple formats incorporating a variety of search criteria based on their needs.

Help Desk Services

Here are a few companies that provide outsourcing or consulting services to the companies who hire them.

ActionCall

Corporate Headquarters:
10900 Wilshire Blvd. Suite 521
Los Angeles, CA 90024
Phone:
800-354-0040
Description of services:
ActionCall offers technical support to companies who hire them. The company issues phone cards with personal identification numbers that give users in need of PC support unlimited toll-free access to the ActionCall National Help Desk.

Amdahl Corporation

Corporate Headquarters:
1250 East Arques Ave.
Sunnyvale, California 94088-3470
Phone:
800-538-8460
408-746-6000

Fax:
408-733-2377
E-mail:
dym00@amail.amdahl.com
Description of services:
Work with customers to assist them in identifying and documenting their information management requirements. Based on the customer's information management requirements, assist clients in improving their existing technology and/or selecting a new technology that is best suited to their environment.

The Bentley Company

Corporate Headquarters:
5 Kane Industrial Drive
Hudson, MA 01749

Phone:
508-562-4200

Fax:
508-568-9468

Description of services:
Offer consulting services for customer service and support including service automation consulting, outsourcing consulting, Internet consulting and survey programs.

The Bultema Company

Corporate Headquarters:
212 N. Washington
Monument, Colorado 80132

Phone:
719-488-9088

Fax:
719-488-9089

Description of services:
Work with companies (especially in areas of service, technology and high-tech support) to develop business, product and marketing strategies.

The Bultema Company is chairing the Support Standards Working Group under the Desktop Management Task Force.

SERVICE 800, Inc.

Corporate Headquarters:
4Q7 East Lake, Suite 200
P.O Box 900
Minneapolis, MN 55391-0900

Phone:
800-475-3747
612-475-3747

Fax:
612-475-3773

E-mail:
jmb@service800.com or
service800@aol.com

Description of services:
While SERVICE 800 is not company to which service organizations can out-source technical support, they do perform a service that Help Desks should know about.

Since 1989, SERVICE 800 has been providing resources to help service organizations follow up with their customers within hours or days of service contacts. Service organizations use the real time, timely follow-ups to impress customers and to collect feedback that measures the effectiveness of individual technicians, products and processes. SERVICE 800 is absorbing follow-up activity from established Help Desks because it is more objective, more timely and less expensive than internal follow-ups.

Because of its relationship with CompTIA (Computing Technologies Industry Association), some SERVICE 800 follow-up programs include benchmarking of monthly service levels against Industry peer groups.

Softbank Services Group

Corporate Headquarters:
699 Hertel Avenue
Buffalo, NY 14207
Phone:
716-871-6400
800-688-4450
Fax:
716-871-6404
E-mail:
sbservice.com

Services Description:
Softbank Service Group provides technical support to over 100 clients including Intuit and Microsoft, in which cases they use help desk software from Astea to build a system to give users e-mail access to its help desk operations. Users can log a case into the system which searches for information to answer the query and notifies the user via an Internet message.

Stream International

Corporate Headquarters:
105 Rosemont Road
Westwood, MA 02190
Phone:
800-507-0363
617-751-1072
Fax:
617-751-7718
Web:
www.stream.com

Description of services:
Providers of over-the-phone technical support for Fortune 1000 companies. The company formed in April of 1995 though the merger of Corporate Software and the Global Software Services Business unit of R.R. Donnelley & Sons Company. They also manufacturer, replicate and resell software, offering more than 40,000 titles.

Sykes Enterprises Inc.

Corporate Headquarters:
100 N. Tampa Street Suite 3900
Tampa, FL 33602
Phone:
813-274-1000
Fax:
813-273-0148

Description of services:
SEI provides support services to help companies achieve service requirements. They operate six customer support centers that provide third-party
hardware and software support. They also provide data processing consulting, programming, application development and systems operations and
maintenance services to clients.

Trade Associations, Research Groups & Industry Networks

The Association of Support Professionals

Corporate Headquarters:
17 Main Street
Watertown, MA 02172
Phone:
617-924-3944
Fax:
617-924-7288

Description of Services:
A software support organization with a network of local chapters that host discussion meetings, social event workshops and facility tours. Members receive research reports, a bi-monthly newsletter, a memberships directory, etc. They also sponsor the annual OpCon Customer Service & Support Conference targeted to the personal computer support industry.

The Customer Support Consortium

Corporate Headquarters:
1601 Fifth Avenue, Suite 1900
Seattle, WA 98101
Phone:
206-622-5200
Fax:
206-292-1825

Description of Services:
With an advisory council composed of professionals at Novell, Dell, IBM, GE Capital, 3M Health Information Systems and Microsoft, the CSC provides a working forum for the technology industry. They are focused on defining industry strategies, standards and technologies that will improve the future of support.

Members receive white papers, executive briefings and analysis documents. They provide workshops and on-line communication forums for announcements and discussions and an annual conference.

Dataquest

Corporate Headquarters:
251 River Oaks Parkway
San Jose, CA 95134
Phone:
408-468-8000
Fax:
408-954-1780
Web:
www.dataquest.com

Description of Services:
Dataquest is a 25-year-old global market research and consulting company serving the high technology and financial communities. They offer market intelligence concerning over 25 specific IT markets. They are owned by The Gartner Group.

The Desktop Management Task Force

Corporate Headquarters:
2111 Northeast 25th Street, JF2-51
Hillsboro, OR 97124
Phone:
503-264-9300
Fax:
503-264-9027
Web:
www.dmtf.org

Description of Services:
The DMTF is a vendor-neutral, not-for-profit organization dedicated to advancing industry standards that improve the manageability of IT and thereby reduce the total cost of ownership. Their goal is to provide a common framework for PCs and encourage vendors to quickly bring manageable products to market.

The DTMF is led by a steering committee that includes Compaq, Dell, DEC, HP, IBM, Intel, Microsoft, NEC, Novell, Santa Cruz Operation, SunSoft and Symantec. They invite participation from industry vendors and technology users.

The Gartner Group

Corporate Headquarters:
56 Top Gallant Road
P.O. Box 10212
Stamford, CT 06904-2212

Phone:
203-964-0096

Fax:
203-316-6490

Web:
inquiry@gartner.com

Description of Services:

The Gartner Group provides research, analysis and advice on IT strategies for users, purchasers and vendors of IT products and services.

Their subscription-based products and services include qualitative research and analysis on trends and developments, quantitative market research and benchmarking services. They also provide consulting services, technology-based training products, worldwide conferences and events, research reports and newsletters. In late 1995 they acquired Dataquest.

The Help Desk Institute

Corporate Headquarters:
1755 Telstar Drive, Suite 101
Colorado Springs, CO 80920

Phone:
719-531-5138
800-528-4250

Fax:
719-528-4250

Web:
www.helpdeskinst.com

Description of Services:

The Help Desk Institute is the largest trade association for help desk pro-fessionals. Founded in 1989, HDI publishes research reports, educational materials and provides training.

Members receive the Help Desk Salary Survey, Customer Support Practices Report, the Help Desk Handbook, Buyer's Guide, an annual sub-scription to their publication and more.

HDI also puts on the Support Service Expo twice a year, which draws thou-sands of exhibitors to attend seminars and see the latest automated sup-port systems.

International Data Corp.

Corporate Headquarters:
5 Speen Street
Framingham, MA 01701
Phone:
508-935-4094
Fax:
508-935-4168
E-mail:
nsethi@idcresearch.com

Description of Services:
Industry expertise in the help desk market as a source for market research.

The Software Support Professionals Association

Corporate Headquarters:
11858 Bernardo Plaza Court, Suite 101C
San Diego, CA 92128
Phone:
619-674-4864
Fax:
619-674-1192
Web:
www.sspa-online.com

Description of Services:
The SSPA, founded in 1989, provides a forum where service and support professionals in the software industry can network with their peers.

They offer specialized training, conferences, publications, special   events and educational programs and services

➲ Chapter 5: Using Other Systems

Outside of call tracking and problem resolution software, there are many other types of automated systems that can make your help desk staff more productive, freeing them up to handle other tasks. An internal help desk may use an automated system to track and manage assets and changes, for example, if responsibility for these functions fall to them.

However, in most medium to large companies, it is not the responsibility of the help desk manager to ensure that everyone in the corporation is using registered software or to order new equipment for a new employee. Still the responsibility falls to IT and the help desk falls under the IT umbrella. The same company you purchase help desk software from often makes software for asset management, change management, tracking inventory, tracking software bugs, etc.

On the external support side some of the companies that make help desk software also make software for dispatching and managing field service technicians, managing support contracts and even software for automating sales functions.

Here's a little bit of information about some technologies that may be right for your help desk.

Remote Control Access

If your help desk technicians are making too many field calls when their physical presence is not required (as it would be when a part has to be delivered and installed), you're wasting time and money.

In many cases (although it depends on the nature of your business and computer literacy of those you support) if support reps could take over end-users' PCs they could solve their PC problems remotely without either party leaving their workstations.

Since most of your end-users who require support are not likely to be technical wizards (otherwise they probably wouldn't need assistance) it can be hard for them to use the correct terms to describe to a tech what's wrong. It only makes sense to let reps look at the end-user's PC for themselves. Your techs can look through configuration files to see why the end-user can't access the database or load a program, for example, while sitting at their desks.

When you can avoid sending help desk personnel out of help desk, even if the help desk is in the same building as the end-user, it will save travel time. Not to mention time saved by not having your reps stopped along the way by other end-users asking questions or needing on-the-spot assistance.

When reps leave the help desk it's also disruptive to service levels since fewer people are around to handle the e-mail requests or phone calls.

The tools available for remote access are also relatively inexpensive, designed with the help desk in mind and some offer features beyond that of a non-specialized program like PC Anywhere. These tools can even be used for training purposes for interactive training sessions involving several users, where less experienced reps, for example, can remotely watch a second level analyst work on a problem as he or she remotely takes over the end-user PC.

The following profile demonstrates how one help desk uses remote access.

Military Support

Stricom is a division of the US army-in-command based in Orlando, FL. They have more than 800 army, navy, government and contract employees who work on PCs across a LAN. They use warfare simulation software, among other software applications.

About 400 employees each week require support. Stricom outsources their technical help desk functions to Sherikon, a professional service bureau headquartered in Chantilly, VA, specializing in engineering and technical services. The help desk has six support reps on the help desk and four technicians who handle escalated problem tickets.

"The help desk solves over 80% of calls without escalation," says Sherikon's Jim Clayton. "The problem was that at least 50% of these calls had to be solved desk-side." The reason for visiting the customer was to demonstrate how to fix the problem since the technical nature of most of these problems makes it difficult for the end-user to describe the situation.

Although the Sherikon help desk is on site at Stricom, Stricom employees work in three large buildings. Support reps often spent ten minutes or more just getting to the end-user's office. Then, after correcting the customer's problem, the rep would often be stopped by another employee with a problem, further delaying him or her from getting back to the help desk. "If all six reps were desk-side, people who called would get an answering machine," says Clayton. "That's not something a customer wants to hear."

To solve this problem they decided to purchase Live Help from Fujitsu (Santa Clara, CA). The system lets them take control of the end-user's PC remotely. "Now we can remotely monitor the situation and can demonstrate how to fix a problem without leaving the help desk," says Clayton. "This makes us more productive, saving us hours. We'd love to have 100 reps available but that's just not practical. That's why we need to maximize the time we have."

Clayton says they choose the Fujitsu system mainly because, unlike other systems, it did not use a TSR (terminate and stay resident) program and therefore did not take up a large amount of computer memory. "We need all the PC resources we can get," says Clayton. For problem management and call tracking they're using help desk software from Applix (Westboro, MA). The system features diagnostic technologies like automatic asset discovery, inventory management and caller history.

Once Sherikon becomes more proficient with Live Help (after further use) they expect to further decrease the need for reps to leave the help desk to go desk-side.

BUG TRACKING

Any organization that employs software developers should be using an automated system to track the stages of product defects. This process is also referred to as bug tracking. When several developers are testing software, especially in a large organization, there needs to information sharing in order to influence progress.

No more than one developer should encounter the same "bug." And the developer who finds this "bug" must make it known that he or she is working on finding a resolution. Otherwise efforts may be duplicated. Once the defect has been fixed, everyone involved in testing the software should be privy to that knowledge.

Tips for Buying Problem Resolution Tools

Buying the correct problem resolution tool involves thoroughly evaluating your company and the vendor. Once you have thoroughly evaluated the processes surrounding the need for a problem resolution tool, a detailed list of requirements should be created. These requirements should be utilized to evaluate the numerous vendor packages. While evaluating the vendors, three areas should be considered: the company, the vendor's customers, and the technology used to create the problem resolution tool.

Vendor companies should be evaluated based upon longevity in the market, their financial strength, quality of the product (ISO certified), their support & training capabilities, and the focus of their product. Their customers should be in similar industries as yours, and most of all, they should be referenceable. Finally, the technology used to create the problem resolution tool should be flexible, the correct functionality for your needs, and a staff of consultants should be available to assist in implementing the technology. Your organization will change and expand over time — so should the package that is chosen.

— Metrix

Here are a couple of organizations that have run in to some challenges, but found quick fixes.

Rolm's High Tech Changes

Telecom equipment vendor Siemens Rolm has 1,600 software engineers who work out of three locations: Santa Clara, CA, Boca Raton, FL and Munich, Germany. Rolm needed a way for those far-flung people to communicate online in real-time.

The software engineer's responsibility, in addition to software development, is to check out all reported problems with software bugs and update the system when the software has been fixed and tested. The software they develop operates Rolm's worldwide installations of PBX systems, ACDs and Phone Mail. It must be bug free as soon as possible.

Their home-grown call tracking system could not synchronize the debugging efforts of all of these engineers working in three different time zones. "The old system had a one day delay before it was updated between Europe and the States," says Rolm's Herman Gepraegs. This led to duplication and wasted time.

The system they chose was Scopus Technology's QualityTeam. They now have faster and more accurate bug tracking and faster problem resolution.

Gepraegs says the customization features were the biggest reason they chose Scopus. "If you look at the features of the Scopus system in general and then look at the way we've customized it, it will look very different," he says.

"We had a mainframe-based system and had to migrate it to a Scopus-based system," says Gepraegs. "The process involved tailoring our whole company around it." Now Unix-based SPARCstations and HP 9000 model workstations connect to each other through a global Wide Area Network.

Adds Gepraegs, "With a commercially available system it's much easier to buy additional tools." They're also using Sybase's database replication server which he says simplifies things. The two databases (one in Munich and one in Santa Clara) always look the same. When data is entered in one server, it's automatically replicated to the other.

Snapshot: Effective Bug Tracking

Company: Advanced Computing Systems

Goal: Since they have close to 40 developers writing software each day, they needed a system that would track bugs and their resolutions. Their old system was becoming overburdened and could not send automatic e-mail messages.

Technology of Note: CustomerQ from Quintus (Mountain View, CA).

Result: Now when employees discover defects they electronically enter it into the system with a resolution, if there is one. Customers involved in product development can also use the system to report defects the same way internal developers do. Their development team can now concentrate on producing software and not have to worry about administrative functions.

Automatic Call Distributors

Most help desks cannot afford the luxury of simply picking up the phone when it rings. It might be fine if you are an extremely small help desk with only one or two people on staff and only receive a few calls an hour.

Help desks with larger call volumes that spike higher on some days need a way to handle and queue those calls. If you've ever had ten calls come in within five minutes without enough people to answer them, I'm sure you can relate. If not enough phone lines are set up, callers will get a busy signal, not a good sign for a top-notch help desk.

An automatic call distributor can route the first call in queue to the next available support rep. We could write a whole book on ACDs, (in fact, I think one already exists) but the definition and description that follows should give you enough information to decide if this is a system you really need to consider.

AUTOMATIC CALL DISTRIBUTOR (ACD) — An ACD answers a call and puts the call in a pre-specified order in a line of waiting calls. On the simplest level, it makes sure the first call to arrive is the first call answered. It delivers calls to agents in a pre-specified order. It delivers the call to the agent who has been free (or idle) the longest or to the next agent that

Smooth Sailing Tips

The best advice I can give someone who has responsibility for a support operation is: define and track meaningful measurements. Help desk applications tend to collect a wealth of interesting and valuable data, but too few operations really exploit that data for the benefit of improving products, service, and customer satisfaction (an entire book could be devoted to this subject).

Our experience shows that virtually 100% of customers who buy help desk applications will implement call logging and tracking processes with their help desk software - its pretty easy. A smaller percentage, between 60-80% will also use their application to build a knowledge base - it's not as easy to do as logging calls. Finally, even fewer, I'd estimate less than half, get serious about using the data they capture to their benefit.

All the vendor products have some sort of reporting capability, and everyone who buys help desk automation has every intention of using it. Too few get this far in their implementation, however. Until measurements have been defined, are being tracked, and actions are being taken to improve them, you should not consider your implementation to be complete. Examples of these measurements include: responsiveness, product defects, workload distribution and others.

—Teubner & Associates

becomes available in a call center. It also provides the means to specify the many possible variations in the order of calls and agents. Last but not least, it provides detailed reports on every aspect of the call transaction, including how many calls were connected to the system, how many calls reached agent, how long the longest call waited for an agent, the average length of each call and many more.

Once the term "ACD" meant a very specific type of telephone switch. It was a switch with highly specialized features and particularly robust call processing capabilities that served at least 100 stations (or extensions). It was purchased mostly by airlines for their reservations centers and large catalogs for their order centers. Companies with less specialized needs bought different technologies that didn't offer the same specialized features. Today true ACD functionality is found in telephone switches that range widely in size and sophistication.

Today there are PC-based ACDs, key systems with ACD functions, key systems that integrate with a computer and software to create a full-featured ACD, PBXs with ACD functions, PBXs with ACD functions that are so sophisticated they compete with standalone ACD systems, standalone ACDs that serve centers with less than 30 agents, traditional standalone ACDs (don't misunderstand, these are usually the most sophisticated), ACDs that integrate with other call center technologies, and nationwide networks of ACDs that act as a single switch.

There is simply no technology more suited to routing a large number of inbound calls to a large number of people than an ACD. Using an ACD assures your callers are answered as quickly as possible. It can provide special service for special customers. ACDs are capable of handling calls at a rate and volume far beyond human capabilities, and in fact, beyond the

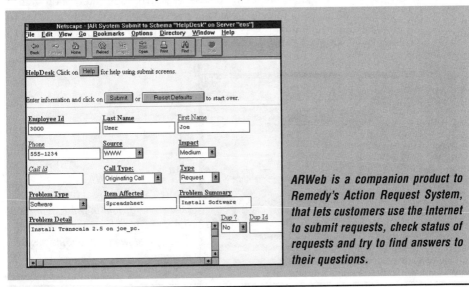

ARWeb is a companion product to Remedy's Action Request System, that lets customers use the Internet to submit requests, check status of requests and try to find answers to their questions.

capabilities of other telecom switches. They provide a huge amount of call processing horsepower. Using an ACD assures your human resources are used as effectively as possible. It even lets you create your own definition of effectiveness. An ACD gives you the resources to manage the many parts of your call center, from telephone trunks to agent stations to calls and callers to your agents and staff.

If ACDs are so great, why does anyone bother with a general business telephone system (such as a PBX or key system)? The ACD's special features are not for every business. Most important, all that processing power and all those special features come at a price. Expect to spend more "per seat" on an ACD than you would for a PBX or key system without sophisticated ACD features. The more robust and feature-rich the ACD, the better it is for a call center, but the more expensive it is. The processing power alone quickly becomes overkill for the average business. (We've seen calculations that say 30 ACD positions use the same processing power as 100 PBX positions.) The average ACD (although certainly not the cutting-edge ACDs) lacks basic business telecom features. They are sacrificed to give more processing power to the tasks at hand. Making an outbound call is a special event on a stereotypical ACD. (To confuse matters there are ACDs that have predictive dialing functions.) Simple general business telephone functions like picking up the phone and dialing someone in another department can be complicated on an ACD.

The plain-vanilla ACD described in the opening paragraph of this definition isn't too technologically impressive these days. (Although it certainly was over 20 years ago when Rockwell first introduced the technology.) Let's follow a call to explore some of the features you can expect to find these days. First, your call will be greeted by an announcement of some kind. The announcement may simply tell you all agents are busy, please wait. Through integration with a reporting system, it may tell you how long the wait will be. It may ask you to enter your telephone number, account number or ID. The ACD will use this to do a database lookup and present information about you (usually on an inte-

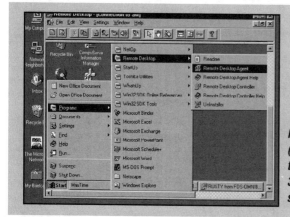

McAfee's Remote Desktop 32 (shown here) lets network administrators remotely manage distributed 32-bit Windows NT and Windows 95 systems.

grated computer system) along with the call (SCREEN POP). The announcement may offer you the option of trying to solve your problem through an IVR system, without losing your place in the telephone queue.

The ACD may also gather information about you through automatic number identification (ANI). It may route your call based on the toll-free number you dialed. A single call center may answer hundreds of telephone numbers. Collecting the dialed number identification service (DNIS) info, the ACD directs your call to the correct agent group (customer service, toasters) or even pinpoints a particular agent that can best help you (SKILLS-BASED ROUTING). The ACD's routing scheme can be configured to distribute your call based on a huge range of criteria, including time of day, day of week, volume of calls in the center, number of agents available and many more. ACDs can put VIPs in a special queue, transfer a call on an order line to the collections department based on the customer ID, send a caller to an agent who speaks her language or send a caller to an agent or group which specializes in the product she needs help with.

While the call is processed, the ACD can provide a supervisor with real-time information about calls in the center, a group or the status of a single agent. It allows supervisors to listen in on calls to evaluate agents, or to join in to assist an agent who is having trouble. Call center statistics displayed on an agent's telephone or on a READERBOARD help the agent manage her own time. Integration with a computer system offers a variety of special features that lace the agents' service software with the telephone system.

After the call, the ACD can automatically give the agent a certain amount of time to finish the transaction before feeding the agent the next call. Statistics for the call are added to the reporting system. Through integration with a computer system, all sorts of activities can be generated automatically from the call — from fulfillment to database updates to the scheduling of a callback.

This definition was meant only to give you a glimpse at some of the functions and capabilities of an ACD. Stay tuned to CALL CENTER Magazine (Subscriptions: 800-677-3435/215-355-2886) for the latest features and technological advances.— *from The Call Center Dictionary (available from Flatiron Publishing, 800-LIBRARY)*

The following stories profile help desks that are using ACD technology.

Dell Computer

Dell Computer, headquartered in Austin, TX is a large manufacturer of personal computer systems, laptops and servers. They mostly sell direct to corporations and home users through their catalogs, but also have some OEMs and VARs.

In fiscal year 1996, Dell's sales reached a record $5.3 billion, up 52% from the year before. They receive about 50,000 calls each day fielded by 1,800 sales agents, customer service agents and technical support staff.

Calls come in on an Aspect (San Jose, CA) Call Center ACD and every call is greeted by Aspect's voice messaging system (part of the ACD) which takes callers through a few menu selections so they can route themselves to the right department.

They have more than 1,200 toll-free numbers, and with every call they get ANI so agents can be sent screen pops. The Aspect system also lets them do skills-based routing so the right call can be matched with an agent of an appropriate skill level.

To staff and predict call volume, they developed their own system. Dell's manager of ACD systems, Mike Nichols, says this internal system predicts call volume and tells them the staff they'll need with a 99% accuracy rate.

"We can predict from our sales the number of technical support calls we will receive," says Nichols.

They recently upgraded to Aspect's 6.0 version. This solved the problem of transferring voice mail messages. Nichols says since they are using four Aspect systems and two PBXs, they were previously unable to transfer voice mail messages to other departments. With the new version they're also now able to monitor agents remotely and can see the reason why agents are logged off the ACD.

The new release also uses Non-Facility Associated Signaling (NFAS), which reduces the number of data channels they need, saving on recurring ser-

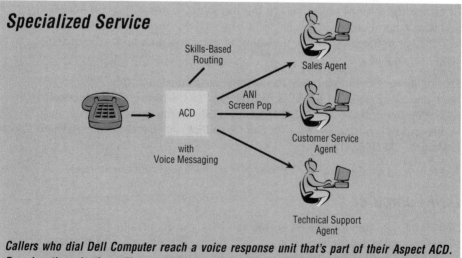

Specialized Service

Skills-Based Routing

Sales Agent

ANI
Screen Pop

ACD

Customer Service
Agent

with
Voice Messaging

Technical Support
Agent

Callers who dial Dell Computer reach a voice response unit that's part of their Aspect ACD. Based on the selection callers make, the they are transferred to the department they need and reach an agent who is specialized in handling the caller's situation (skills-based routing). Agents are sent a screen with the caller's file along with the call.

vice fees and future installation charges it pays long distance carriers for signaling. It also makes more channels available for customer calls.

Jostens Learning Center

Jostens Learning Center makes integrated learning systems used in schools for teachers to deliver lesson plans and students to do the lessons on computer. Their systems are also used in prisons, and in GED training programs to teach life skills, and used for assessment testing.

At their San Diego, CA support center they take questions, mostly from teachers and lab technicians, who need installation, integration and other kinds of usage information.

Calls to the support center go through a Telcom Technologies (Pomona, CA) ACD 6000 switch through which they've set up voice prompts. Callers are prompted to make IVR selections based on the product they are calling about. They can also opt out of the system to reach a live voice. Field technicians also use the system to leave messages.

"Our goal is to answer 80% of our calls in under a minute," says Jostens' Russ Gangloff. "These teachers are calling from a classroom with screaming kids and need to be up and running."

Using off-the-shelf software they can do key word searches to narrow down problem descriptions. They've also scanned in technical manuals and documentation that techs can view on their computers.

IVR Systems

An interactive voice response system is an ideal solution for help desks looking to ease the burden on support reps. IVR can reduce abandoned calls, keep your help desk open 24 hours a day, decrease support costs — and give callers control.

Smooth Sailing Tips

Most, if not all, problems, stem from basic management deficiencies. Most of these are in the people management area. The wrong type of person is hired - typically they are very technical but can't follow the rules and can't communicate with customers. Another typical problem is that the rep really wants to do some other job and thus doesn't take the support job seriously. This attitude is often reinforced by a company atmosphere that makes it clear that support is not important and/or customers are pests. Often CSRs are not given proper incentives, i.e., finishing "special projects" counts more than taking calls. One of the main benefits of a request tracking system is that it allows proper rep incentive plans to be put in place.

Occasionally the problem is a tool and technology problem. These problems are typically associated with home-grown help desk tools or old monolithic mainframe based products.

— Opis

An IVR system has many functions in the help desk. It cuts costs by cutting the time techs spend on the phones. An IVR system (depending on its use) can pay for itself in less then a year. Since it can handle and complete calls without human intervention, you're likely to need fewer support reps, or support reps can spend time handling other help desk functions.

When creating applications, it's very important to create a system your callers will like. A problem with some IVR applications is the setup. Giving callers too many choices only creates confusion and defeats the whole purpose of IVR. For example, you don't want to create a situation where callers are constantly pressing zero to reach someone live because they couldn't remember what to selections to make. More than five choices can be confusing.

After callers make a first selection — say they choose software support for example — then it's okay to have four or five more choices so they can indicate the type of software they need help with. It's even okay to have a third tier asking, for example, what version of Windows they need assistance with.

The last thing you want to do is frustrate callers with an IVR system. They will like the system if it can give them the answers they need more quickly than waiting on hold for a rep and being repeatedly transferred. But during business hours callers should always be able to press zero for an operator or hit another key to get live support. It's very frustrating to be locked into a system. After hours it's acceptable as long as callers know live support is available during business hours.

How it Fits in the Help Desk

You can use IVR to handle many tasks, from the simplest to the most complex. Here are some examples of how you can substitute an IVR system for valuable rep time.

- An IVR system can prompt callers to enter an identification number that will search the database to check if the customer is under warranty. If not, it can tell the caller what action to take. This helps shorten the call if a support rep typically checks warranty information before handling a call.

Tips for Buying Problem Resolution Tools

The most important requirements for call center and help desk systems are the following:

Ease of Use — support solutions must offer near-total "out of the box" functionality.

Scalability — systems must be able to grow with the company and handle increased numbers of users seamlessly.

Flexibility — support software must be adaptable to your existing processes and to the ways you do business.

Customizable — businesses must be able to change the product to serve new processes and new business models.

— Scopus Technology

- If you are an internal help desk, you may get a lot of calls from employees who need their terminals or printers reset. Here you can use an IVR system to prompt callers to enter their terminal ID number. The system can then get the network information from the database and advise callers of the action to take to reset terminals or printers.

- Use it as an alternative to waiting in queue. An IVR system will give callers options other than holding in queue or finding a busy signal. Many help desks really don't staff to meet call volumes. IVR can help to lessen the burden during busy times.

- Most impressive, an IVR system can let callers troubleshoot. By combining an expert system with an IVR system you can have the system ask callers questions to try and identify the problem the same way a support rep would.

There are a couple of vendors who actually combine a rules-based expert system with an IVR system in one package. Such a system, like The IntelliSystem from Intellisystems, (Reno, NV) for example, will typically start at $25,000 (depending on configuration) for four ports.

This is no doubt the most effective use of IVR, saving customers (and the help desk) time. Customers are also likely to feel empowered when they can solve their own problems without relying on anyone.

What it's done for them

One problem I hear from many help desks is that they have a tough time convincing upper management that the expense of automating is justified. That was a problem at Hewlett-Packard. So they decided to ask customers how they felt about automation. Callers go through their IVR system to retrieve the information they need.

Then the system asked callers if they would like to answer a survey by letting the system ask them questions. Callers indicated responses to survey questions by pressing appropriate keys on their phone pads. The information was captured in the database, and this data gave HP feedback that helped them show other departments and management why they needed to automate.

At Novell, callers enter an incident number to be transferred back to the support rep who is working on their problem. This eliminates the need for support reps having to ask for an incident number or ask the caller to repeat his problem.

At Quantum, one of the world's largest suppliers of hard disk drives, they use IVR to give callers a choice to use automation when there are long waits in queue. Callers can select to speak with a rep, receive documentation via fax-on-demand, use their bulletin board service or request a manual, all through the IVR system.

Automating (using IVR and having reps use problem management software) has made for a great return on investment. Just three years ago Quantum was only handling 6,000 calls a month but had 24 technicians. Today they can handle four times the number of calls with a third less staff.

IVR transactions are stored in a database so you'll know the areas in which callers are seeking help. It's a good way to track if too many calls are coming in on a particular topic. It can be an indication of a problem with your product.

You can examine callers' IVR selections to determine what options callers request most. Naturally if callers most often select fax-on-demand, that should be one of the first menu items.

Build or Buy?

You may wonder whether or not you should just purchase the tools to build you own applications or purchase a complete package from a vendor. If you have in-house developers you can use application generator software to build the IVR applications yourself. You'll save money upfront and if you need to make changes to the application, you won't need to rely on the vendor.

But for problems, you won't have vendor support. You'll need to have the developer on hand. Figure under two grand for a four-line card kit from one of the board manufacturers. Then just add the developers' tool kit, write a little software and you'll have a full fledged IVR system.

Computer Telephony Integration

Information about callers who dial your help desk lies in your computer. Reps must find out who the caller is, perhaps type in an identification num-

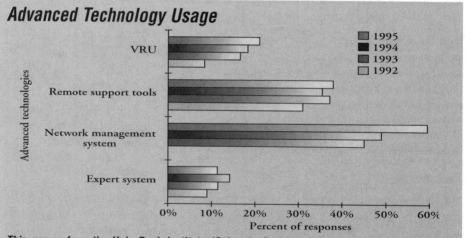

Advanced Technology Usage

This survey from the Help Desk Institute (Colorado Springs, CO) shows how under-utilized voice response units (VRU/IVR) are in the help desk (although their use is gradually increasing). HDI terms these four tools as "advanced technologies." On the other hand, almost 90% of respondents said they were using "core technologies" like voice mail and e-mail.

ber to bring up information for each caller. When you use computer telephony integration, reps have all the information they need about the caller before they even say hello.

CTI applications require the linking of your telephone with your computer to apply computer intelligence to telecommunication devices like switches and phones. Buying some middleware that can make your phone and computer understand one another can make it possible for reps to receive a profile of the caller with every call.

For a help desk application you could set up your IVR system to ask customers to indicate their contract or identification number when calling for support. Based on the number they enter, the system could transfer the call along with the caller's record, pulled from the database, to the appropriate support rep. Or you can program your switch to send the caller's phone number (using automatic number identification or Caller ID) to the middleware.

Using CTI saves the rep from having to ask the caller his or her name or for an identification number to find out who they are and whether or not they have a valid service contract, the equipment they're using and problems they may have called with in the past. Not having to search for this information can save up to 20 seconds per call, adding up to significant savings for any company handling a large volume of calls.

Take Novell, for example. Reps tell customers whose problem can't be solved on the first call, to call back later to find out if their problem has been solved. When callers reach the IVR system, they are prompted to enter their incident number so their call can be routed back to the rep working on the problem. The rep gets a screen full of information with every call (also known as a "screen pop").

You can now use CTI for Internet applications, linking your help desk's phone system to the Internet. For example, you may have a dedicated Web site set up so callers who use your knowledge base or other reference materials can try to find their own solution. Typically, if they are unsuccessful, they will terminate the connection and call the help desk or send an e-mail request. There are now companies that provide applications which let the customer reach the help desk by clicking on their mouse to set up a phone call to the help desk.

Smooth Sailing Tips

It would be helpful for help desks to be categorized by type of call and call volume. The support lines should be divided by the specialist's expertise and experts should train the first line specialists so that they can handle the more difficult calls in the future. It is important to have training tools available so that help desk specialists can get the proper training required.
— Fujitsu Software

To find out more about CTI technologies and where you can find manufacturers of this technology for small and large applications, consult Computer Telephony and Call Center Magazines. For more information call 212-691-8215.

ISDN

ISDN (Integrated Services Digital Network) is a collection of telecommunications transmission standards and services with a goal to provide a single international standard for voice, data and signaling (so one network can be used for all three purposes, instead of having three networks); make all circuits end-to-end digital; use out-of-band signaling; and bring a significant amount of bandwidth to the desktop.

What can ISDN do for you? Here are some examples:

Selective Call Screening — Lets you know who is calling before one of the phones is answered. You can answer calls from high-priority customers first. A similar feature tells you who is on the phone even if the line is busy, letting you further prioritize calls in the center.

Shared Screen — Switched data services provided via ISDN let two people in remote locations, both equipped with a computer terminal, to view the same information on their screens and discuss its contents while making changes — all over one telephone line. Great for technical support centers. Have your techs confer with experts all over the country, or simply take a look at the customer's problem.

Network Access — Lets your reps easily access one of your company's databases, even if they don't use it often enough to be on "the network."

Less Down Time/Cost Savings Moves — When a company moves an employee within an office, there can be hours or days of lost production while a computer terminal and phone set are being installed. In some cases, the terminal is connected to a network via coaxial cable. ISDN virtually eliminates down time, as well as the need for coaxial cable.

A key component of ISDN is CCITT Signaling System 7. This is an international telecommunications standard that does two basic things: First, it removes all phone signaling from the present network onto a separate packet switched data network, providing more bandwidth. Second, it broadens the information that is generated by a call, or call attempt. This information — like the phone number of the person who's calling — will significantly broaden the number of useful new services the ISDN telephone network of tomorrow will be able to deliver.

ISDN comes in several flavors (or services). They are:

1. The 2B+D "S" interface (also called the "T" interface). The 2B+D is called the Basic Rate Interface (BRI). The "S" interface uses four

unshielded normal telephone wires (two twisted wire pairs) to deliver two "Bearer" 64,000 bits per second channels and one "data" signaling channel of 16,000 bits per second. An S-interfaced phone can be located up to one kilometer from the central office switch driving it.

2. The 2B+D "U" interface. This "U" interface delivers the same two 64 kbps bearer channels and one 16 kbps data channel, except that it uses 2-wires (one pair) and can work at 5 to 10 kilometers from the central office switch driving it. The "U" interface is the most common ISDN interface.

3. The 23B+D or 30B+D. This is called the Primary Rate Interface (PRI). At 23B+D, it is 1.544 megabits per second. At 30B+D, it is 2.048 megabits per second. The first, 23B+D is the standard T-1 line in the U.S. which operates on two pairs. The second 30B+D is the standard T-1 line in Europe, which also operates on two pairs.

4. A standard single line analog phone. A 2500 or a 500 set.

ISDN has been a long time in coming. Some of the frustrations of its unkept promises are reflected in the "other" definitions for ISDN: I Still Don't Know; It Still Does Nothing; Improvements Subscribers Don't Need; I'm Spending Dollars Now. — *from The Call Center Dictionary (available from Flatiron Publishing, 1-800-LIBRARY)*

Pop up windows in Quintus' CustomerQ let users view several tables of information in one window.

➲ Chapter 6: The Self-Help Evolution

Using E-mail

More and more help desks are realizing the value in letting end-users reach them over the Internet or through internal e-mail systems.

End-users no longer need to rely on the phone to reach the help desk. They can use e-mail to submit a problem ticket, check the status of their problem and even search an online knowledge base to try and find solutions to their problems.

The savings to the help desk are enormous: "If someone can help themselves by using a Web site, there's an incidental charge of two or three dollars as opposed to an average of $20 to $25 dollars to handle the same request with live support over the phone," says Tom Sweeny of Dataquest.

Sweeny also says using the Internet has become more widespread thanks to improved search capabilities and companies who now update their Web pages daily.

Predictions are being made today that in the not too distant future all computers will come with help buttons users can click to reach an assortment of information designed to help with specific problems.

Tips for Buying Problem Resolution Tools

In buying a problem resolution tool, customer support managers should look for applications that allow them to take full advantage of the collective knowledge of their support organization. A support organization's analysts are their greatest resource and if the organization can capture and reuse those analysts' knowledge and immediately share it throughout the organization, they can dramatically reduce the time required to solve customer problems.

This is especially crucial for those organizations in the high-technology industry, where products change rapidly and there is continual pressure to solve new and increasingly complex problems. Buyers need to be wary of systems that do not allow them to capture and reuse knowledge in real-time, since such products will hamstring any increase in productivity a problem-resolution tool might otherwise provide.

— Primus

The Internet

Web links have crawled into an astonishing number of help desk systems. But is this an add-on that will sweeten your profits or sour your customers?

The proliferation of Web links showing up in help desk vendors' suites of products has generated a great deal of interest, mixed with some skepticism.

In the self-support revolution, it is now possible to put your problem resolution system and knowledge base on the Internet so your customers can search your knowledge base using either key words, case based reasoning, or some other form of artificial intelligence.

Both internal and external help desks use bulletin boards to post information such as frequently asked questions, lists of top ten problems and their solutions, advice from other PC users, tech notes and manuals. Many help desks also allow employees or customers create a problem ticket through the Internet.

The Web is also a great tool for its marketing ability. No doubt, it makes it

Mary Kay Cosmetics

Mary Kay Cosmetics has offices all over the country. Several thousand independent beauty consultants work for them in addition to 350,000 people who sell the cosmetics.

The external help desk operation that supports their beauty consultants and directors is separate from the internal support provided at their corporate offices. Directors of beauty consultants, for example, are using a new software product so they can monitor performance. They often call the help desk with software operational questions, says Donny Fulton, an independent technology consultant hired by Mary Kay.

At their corporate office in Dallas, TX, they've networked their computers to log calls and fix problems without any human intervention. These computers run everything that Mary Kay does. Fulton says they have about 15 software packages running to perform specialized functions and to support this operation. Such packages include a console managing system and network managing software.

They use help desk software from Silvon Software (formerly Raxco, Westmont, IL) to track calls and route them to the appropriate second level analyst if the system can't solve the problem without human aid.

"The key is that we can monitor systems and catch a problem before it turns into a real problem for the end-user," says Fulton. "When you let the computer do as much work as possible, you can do more support with the same number or even less people." He says it also eliminates the need for a first level support tier.

Fulton says that within a couple of weeks 40 to 45 second level support technicians will be using Helpline. "We were using another help desk package but it couldn't support the volumes we needed and it was a PC-only package. What we liked most about Helpline was the architecture. "Since we have several different operating systems we needed something that could run on multiple platforms.

easy for anyone to navigate at their own pace, just clicking on a mouse to request a manual, or other product literature. All of these uses are great ways to become proactive and decrease the volume of calls to the help desk.

While no one can question the value of the Internet, its use for letting end-users solve their own problems online the same way a support rep would can be overrated in terms of it's success.

In a help desk that staffs its first level of support with people who are not technically oriented, there can be value in letting end-users search a knowledge base rather than calling a low level rep if the system you're putting online uses some form of artificial intelligence.

If your problem resolution tool only lets you use key words to search a knowledge base, you're likely to be staffing your first level with support reps who are either well-trained or at least somewhat technically oriented. That makes the most sense since most experienced technicians find it best to search a knowledge base using key words. Many say they spend less time going this route instead of using a product with built-in artificial intelligence.

Putting your resolution tool online when your end-users may not know what word to use to search the knowledge base to come up with the right solution cannot be as effective. And if customers can't come up with the right solution quickly, they are not going to be convinced to continue using the Internet. Just one bad experience can lead them back to square one — using the phone for support.

Many vendors now make Web modules designed to work in conjunction with

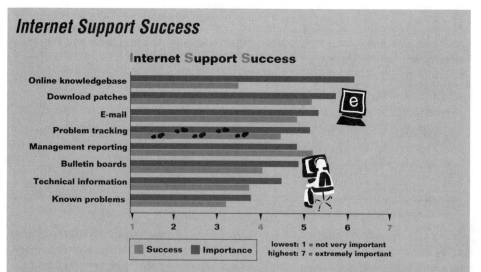

Internet Support Success

Based on their research, International Data Corp (IDC., Framingham, MA) found that success over the Internet does not match the importance of using it for these functions.

their problem resolution software. Such a system lets end-users create a trouble ticket and search a solution database over the Internet. End-users can post a problem and later check on the status of their request if they could not find a solution on their own. Field support staff can also use the system to view calls assigned to them and collect information from a remote location.

A common concern for many help desk managers is whether or not using the Internet could really save customers time and lead them to the information they need, as opposed to picking up the phone to reach a support person.

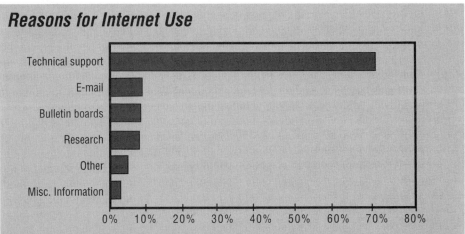

Reasons for Internet Use

Overwhelmingly, when help desks make it possible for end-users to use the Internet, most do so for technical support. source: IDC

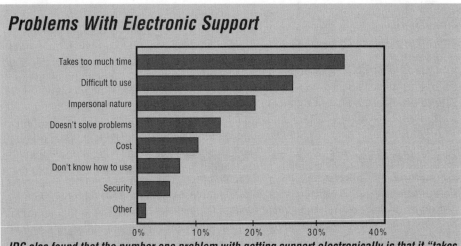

Problems With Electronic Support

IDC also found that the number one problem with getting support electronically is that it "takes too much time." That's expected to change as more tools are developed and the Internet becomes easier to use.

When you look at hold times customers often face to reach a support rep, it's certainly possible that customers could close out their own problems in less time than when waiting for a rep. But no one seems overly anxious to risk spending the money for a Web link if it really won't cut calls to the help desk and only frustrate customers who have negative experiences online.

According to International Data Corp. (Framingham, MA) there are about 80 million users on the Web today, but by the year 2000 IDC expects to see that number grow to 200 million, making the Internet a $5 billion a year industry.

Still, in mid 1996, only 10% of companies were relying on the Internet for business functions. According to IDC, healthcare, finance, banking and education are industries where Internet use should raise that percentage dramatically as it becomes more secure, reliable and as more access tools are developed.

While IDC shows the Web being used most for obtaining customer service and support information, 80% of customers rely on the phone for support questions. Only 19% of customers request support through electronic means.

One reason is that finding information over the Internet can take too long and it can be difficult to find the right information. Plus some people will find it impersonal, preferring to talk to someone about their problem.

For technical support, the Internet is used most for software and hardware support. The three most popular uses for intranets, according to IDC, are (in order of importance): in human resources to give employees information on 401Ks and insurance options; technical support; and to post general company details.

It's important to ensure information posted on a Web site is up-to-date, because

Automatic Terminal Reset

COMPANY: National Semiconductor.

GOAL: To reduce the number of calls to the help desk for callers who need their workstations and terminals reset.

TECHNOLOGY OF NOTE: Edify's (Santa Clara) Electronic Workforce.

RESULTS: Previously users would call to speak with someone at the help desk to find out the terminal reset function. Now one of three specialized Edify automated software agents answers the hotline and prompts callers to enter their terminal ID number.

The software agent then gets network information from the host and informs the caller. Depending on the user's action, the software will disconnect the caller or reset the terminal. The system asks the user to press a key to indicate a successful completion.

The software handles about 110 terminal requests each month allowing technicians to handle more complicated situations. They plan to use the software agents to field other types of calls to the help desk in the future.

a negative online experience affects future use if an unsuccessful incident prevents someone from going online again. Since 20% to 25% of help desk calls are avoidable, the right marketing is important in an effort to attract new users.

Success using an online knowledge base is not high, because when you look at the importance versus the success rate there's still a long way to go. Microsoft is a rare example of a company that receives more support requests electronically than by phone.

Here are some tips for anyone getting started with providing electronic technical support:

• Use the Web as a differentiator to complement, not replace over-the-phone support.

• Understand customer needs. Some people may feel that they would rather speak to a person, but you can demonstrate the value of using the Web.

• Communicate usage and capabilities. Use your hold message as a commu-

Smooth Sailing Tips

Be proactive. Reactionary support is a given. The only way to drive your call numbers down is to systematically eliminate calls. Use the reports generated by your help desk system to review staffing needs, and properly staff your help desk for the loads you have. Review other reports for trends showing areas where equipment could be better utilized, where a certain percentage of calls could be eliminated with an inexpensive piece of software, etc. When the help desk becomes aware of a problem that affects an entire department, they should immediately contact that department with a solution or a backup plan, so the group is not completely incapacitated while the problem is being solved. It also makes the end users feel more confident that the help desk is on top of things. In the same way, immediately notify all the users when a system is back up and can be used normally again.

Train the end-users. Train all your new employees, and make sure that training is implemented when any new system is put in place. This will make your users more self-sufficient, and lower the number of calls to the help desk.

Cross-train your help desk staff. As much as possible, try to get the various "experts" to work with each other so the general level of experience and knowledge gets more evenly distributed. This way the networking calls don't always have to back up because your one network expert is out sick, etc. It will also increase job satisfaction because the analysts will feel more comfortable handling a larger variety of questions and problems, and will feel more empowered.

Try to rotate the help desk workers so they're not always on the phone. Being constantly at the mercy of the users with nothing else to do can be a drag after awhile. Avoid burnout and expand the technical knowledge of the help desk by staffing it so the people on the phones can rotate in and out, and work on projects, developments or changes during their time away from the phone. This helps to increase job satisfaction and minimize turnover.

— Utopia Technology Partners, Inc.

nication vehicle that will read: 'Try our home page at www.XCompany.com.'

The following are three examples of well-known companies that have had great success using the Web for more than technical support:

- Microsoft has dedicated bulletin boards for internal developers. They use information from these electronic bulletin boards to improve customer service. For example, they had viewed bulletin board dialogue among Excel spreadsheet users who were talking about problems with migration in Lotus 123. Since learning of these problems, they responded by creating a function to help with the migration.

They also use the Internet as a marketing tool. For example, they may use it to inform developers of a situation posting a message like: 'here's the biggest problem with Lotus 123. If you can fix it, it will be a better sell.'

- At Digital Equipment Corp. customers can get up-to-date pricing internationally over the Web so they can compare prices in different locations, which can add up to significant savings for a 500 user site, for example. Then they can download the software right over the Internet after purchasing it.

- Through Federal Express' home page, customers can view shipping choices in all areas and see all options. They can track their shipment, see how it's progressing and even see who signed for it.

Breaking A Support Bottleneck
How one company lets customers support themselves.

Company: Shaw's Supermarket, the second largest chain in New England; and Innovax (a company that has installed about 1,000 checkout systems at locations across the country). Innovax supports their own support center operation and provides support to Shaw users and other customers.

Goal: After Innovax sells their point-of-sale systems (called the Aurora Supermarket Application) they need to provide support for these systems. They began using the Apriori Technical Support system from Platinum Technology/Answer Systems years ago for first and second tier support. But more recently they decided they needed to roll the product out through their enterprise so all personnel could access Apriori from their desktop computer.

Innovax's Jim Moscati says Shaw's Supermarket chose the Innovax system because they needed a full-featured open system. "They wanted to use our software on hardware they had from multiple vendors and they were able to do this," says Moscati.

Technology of note: Innovax's (Irving, TX) Aurora Supermarket Application and Platinum Technology/Answer System's (San Jose, CA) Apriori.

Results: Shaw's has a help desk on site but they call Innovax' help desk when they need assistance with complex problems. Innovax has one central help desk that handles calls for all stores that use their product.

But customers don't need to speak to a support rep. Since Innovax provides customers with a modem and communications package with every system, customers can dial into the system to access their knowledge base.

The Apriori client/server help desk software makes this "hands-free" support possible using the Aurora supermarket application via an e-mail connection.

Moscati says that customers find a solution to their problem themselves 60% to 70% of the time. Innovax support personnel can also dial into the stores directly to perform remote diagnostics.

Throughout Innovax's enterprise network, many departments use Apriori. For example, Marketing uses Apriori Mail to post news and events of interest to Innovax customers; Quality Assurance uses it to create problem reports that notify Development of software changes; Accounting uses Apriori to notify Technical Support of the contract status of customers; and the Innovax Distribution Center uses it to track customer equipment returns, to update configurations for current customers and input equipment and software configurations for new customers.

Customer Service At Work

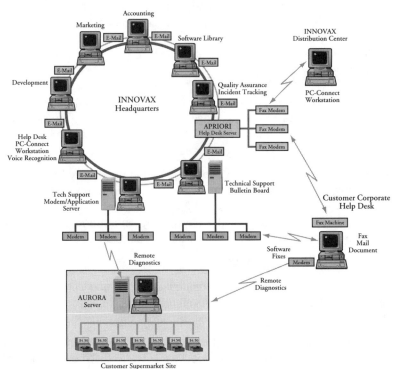

➲ Chapter 7: Managing Your Help Desk

Organization and Maintenance

I f you knew all the secrets to organizing and maintaining a top-notch support center you would probably be one of the most sought out help desk managers. It's really not hard to learn. Everything is in the process. Here are some steps to follow.

1. Assess Technician Skills

Asking technicians to answer questions regarding their areas of expertise will enlighten your whole desk. A technician who does not know any Visual Basic programming, for example, may get a call regarding this program but not realize that someone in the same office is an expert in Visual Basic and better suited to handle the caller.

That person will know not to waste time on the call when there is another person who is an expert and can help. If you let employees take calls in areas where they are most skilled and areas they are most interested in being in, they will be more productive and eager to learn and help others.

2. Be Wary of Metrics

Ralph Tarantino, a help desk manager at McDonald's Corp. questions the logic of relying too much on numbers as an indicator of service levels. "You cannot succeed if you cannot measure, but you need to be careful that you don't get caught up with metrics and measurements," he advises.

Sometimes companies get so focused on measuring that they forget what their deliverables really are. "You can rely too heavily on numbers as an indicator of how well your desk is doing," Tarantino says. "A good support manager needs to be involved in the day to day operations of the desk. They need to be able to get the pulse on the people side of the desk."

3. Centralize Your Database

Tarantino thinks it's important for a help desk to have one phone number, one centralized database, one mission statement and service levels that apply to everyone, even if the help desk has more than one location.

"The help desk database should be shared with other departments like asset management, marketing and the sales force," says Tarantino. "The sales department can look up a customer's file to see what problems they've been having."

Other departments can make themselves look good by helping a customer with a problem if they are aware of it. "Asset management can supply warranty information that can assist the desk in calling the right provider to service equipment," says Tarantino.

If the sales department sees open tickets concerning problems with a particular product, they know not to visit the customer trying to sell anything relating to that product for a while.

4. TQM

Tarantino says Total Quality Management (TQM) has an 85/15 rule: 85% of the time problems are in the process, 15% of the time they are with the individual.

"Help desk managers have a habit of looking to the technician if their abandonment rate is high," he says. "But this can make the person feel like a failure when really it's the process at fault." He says it's almost impossible to achieve 100% customer satisfaction, but it should be a goal.

He says you should involve analysts when considering ways to improve customer service. "If you get their input, it becomes a group decision," says Tarantino. "People are more likely to buy into an idea when they have input."

The help desk should be involved from the beginning stages of development through the roll out of a new product. "You can't just give calls regarding a new product to the desk," says Tarantino.

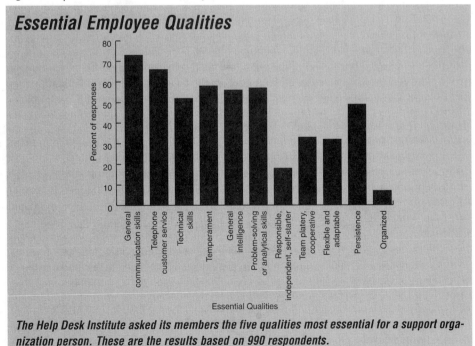

Essential Employee Qualities

The Help Desk Institute asked its members the five qualities most essential for a support organization person. These are the results based on 990 respondents.

"The help desk needs to determine the number of calls adding a new product may generate, the skill sets of agents who could support the product and the additional hours they may need to be on call. When the support analysts work with the developers early in the process, they build a relationship where they can rely on one another in the different phases they will go through."

Tarantino also suggests marketing your help desk. "Go to the customer and tell them how the support process works," he says. "Many customers are surprised to learn the number of calls and types of calls analysts handle every day. They will not be thinking of that when they are holding in queue unless they are told."

Preventing Burnout

Help desk reps certainly don't compete for any awards for job longevity. Turnover among support reps is high. But it's not hard to create a workplace that reps won't want to leave. It's also very cost-efficient.

Employee burnout is a big obstacle to many help desks. Turnover is very high among support staff and it's easy to see why. They spend every day dealing with and trying to solve other people's problems.

Answering calls and staring at a computer screen all day is tedious work. Often, the callers support reps have to deal with are already frustrated with a problem they are having. Combining that with a wait on hold can make for annoyed customers before they even reach the desk.

Yet reps are required to remain polite while trying to get the customer off the phone so they can get to the next call. The impression a support rep leaves on a customer is representative of how they perceive your company as a whole.

It's in the best interest of the help desk to retain good support reps for as long as possible. The longer they spend with the company the more familiar they are with procedures and the more knowledge becomes stored in their heads.

"The number one reason for people leaving the help desk is having no career path within the help desk," says Tarantino. "Any good desk realizes that the talents and expertise of their help desk can be used in other departments or for special projects. A good manager should help promote their people to other areas."

Since it's a lot for one person to manage 10 to 15 analysts, one suggestion is to find a group coordinator in a specialized area, such as networks.

This will give that person some management input and some free time off the phone. Other options include letting some analysts work on metrics or improvement initiatives.

"Support desks should put out a company newsletter listing short term positions in the help desk," says Tarantino. "There may be programmers who want to jump in for six months and a help desk analyst who wants to work in development for a while. It's good PR for the desk."

He says some companies even make new employees (expected to work in other areas) work in the help desk for several months. "It exposes them to the culture and environment of the company," says Tarantino. "And customers get to know them. Even executives can sit at the desk for a few days or a week. It helps them realize how much help desk analysts need to know."

Tarantino suggests team rewards when customer satisfaction is up. He prefers group rewards over individual recognition because the help desk as a whole has to look good.

Free lunches, team outings, time off and meetings where teams get special recognition and awards are all good incentives.

Reducing Turnover

Once you've realized why burnout may be so high among help desk employees, you've fought half the battle. Now you can take a proactive role in making a conscious effort to reduce turnover.

This is a true story: a friend of mine who is employed as a clerk on Wall Street loves going to work every day. He always talks about how much he loves his job. He actually looks forward to going to work every single day and as a result hates to ever be late or take a sick day. And he's been at the same job for six years.

So what's the attraction? He knows that people rely on him. He takes orders from brokers and passes them on to traders. If he makes a mistake, the company can lose a lot of money. This makes him feel like his position really matters and that a big part of the overall transaction rests on him. He feels like he's a part of a team. He often repeats stories of times when he's complemented on his performance. His face literally glows when he tells these stories.

The moral of this story? Rewarding people for performance really makes them feel good about themselves. Even a simple "you're doing a good job" really goes a long way. People who feel good about themselves perform better on the job and have a more positive attitude.

There are several ways to ensure good people stay with you.

1. Have employees fill out surveys that tell about themselves, their experience, preferences and dislikes. You may find that a rep who is handling mostly calls on WordPerfect problems is much better suited to be handling questions pertaining to Excel. Have employees indicate areas where they would like to be solving the most problems. They'll prosper and so will you.

2. Ask employees if they feel like they are part of a team. Have them suggest other job functions that they would like to be responsible for when they are not solving problems.

3. Ask employees how they feel about the people they work with. Have them indicate how they feel about their relationship with managers and if they feel comfortable talking to managers about problems on the job or to make requests to work on special projects.

4. Are there areas where they would like to receive additional training? Perhaps they wish they could learn more in a particular area that would help them in their position. Based on responses, consider holding monthly seminar classes on popular requests. The costs are minimal if you are educating your employees in areas that they want more information about to help on them the job.

5. Ask them what would make their jobs easier. People feel empowered when their feedback counts. Letting employees make decisions goes a long way in making them feel empowered. If an employee feels confident seeing a problem through from start to finish, even if that is not typical corporate policy, it is a real confidence builder.

6. Make sure support reps have opportunities to interact among their peers. Hopefully, you're using an automated help desk system so that when a problem is solved for the first time, the answers are stored in a database for all to share and no one has to solve the same problem again. Information sharing in other ways should also be encouraged. It's a great way for less skilled reps (perhaps first level) to learn from more experienced technicians.

7. When holding a gathering of support reps plan to have about an hour or so of structured activity. Have a couple of technicians speak about experiences over the past month. Have them address some trends everyone should be aware of. Use such a meeting as an opportunity for support reps to come together under informal circumstances. Ordering lunch (sandwiches or pizza) for the crew will automatically create a light atmosphere.

Remember too, that if you are evaluating anything from a piece of software to a chair, you need to get input from staff. They are the ones who will be spending the day using the purchases made for them.

A piece of software that doesn't offer certain customization options may seem minor, but when a person is spending eight hours a day looking at that screen, not being able to make simple changes can be frustrating. Something as simple as a chair that's not adjustable can cause fatigue to a five-foot-one or six-foot-two person. People work best when they are comfortable.

Employees should be made to understand how their work affects the organi-

Tips for Buying Problem Resolution Tools.

The best problem resolution tools enable companies to take advantage of their own experience and information base and let users enter a query in natural language.

— Magic Solutions

zation's bottom line. It's important for staff to know the company's strategy and long-term goals. It helps them understand their relationship to the company as a whole and encourages team building. Sharing vision can make the help desk feel that their job performance controls the success of good support.

When working in a help desk is viewed as a dead-end position there's no reason for employees to strive to be the best. There's no incentive because there is no higher level to be reached. You need to promote support reps, perhaps to managerial or specialist positions.

Celebrate promotions with a small party so everyone sees for themselves that there are places to go for hard workers. It's a perfect opportunity to demonstrate that high achievers don't go unnoticed.

Promoting the Help Desk

Every area in your organization should know what the help desk is doing. You can use newsletters, e-mail, electronic bulletin boards or even monthly or quarterly meetings to keep other departments aware of progress. List the five most popular questions or problems and their solutions. That will educate departments who use the help desk of solutions to some common problems, perhaps eliminating some calls to the help desk.

You should also inform other departments when the help desk is doing a good job. If average speed of answer is up or if you've solved 10% more problems than the previous month, say so.

Upper management cannot consider the help desk a black hole they keep

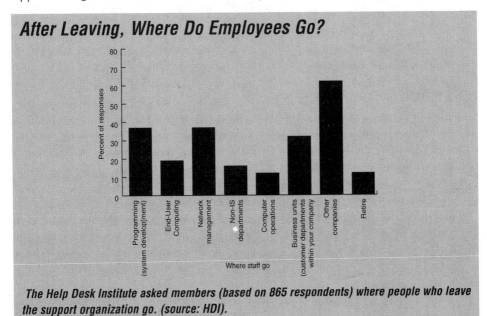

After Leaving, Where Do Employees Go?

The Help Desk Institute asked members (based on 865 respondents) where people who leave the support organization go. (source: HDI).

shoveling money into or the forgotten department when the progress and savings the desk has achieved is spelled out.

End-users also like to hear about improvements in service. If you've recently automated or made other improvements, tell them about new expectations and how they will be affected.

If your company has user group meetings for external customers or internal employees, use these meetings as an opportunity to introduce help desk staff to the end-users they support. This will let end-users see help desk reps as real people, like themselves and put a face to their voices.

Telecommuting

You can reward those who have been reps for two years, for example, by letting them work a day a week from home. If they can tie into your database from their home PC's there's no reason why you shouldn't. They could even serve as second level reps on days at home, checking e-mail to try to solve the problems that couldn't be handled at the first level and require call backs.

Telecommuting is starting to catch on in corporations around the country. The Help Desk Institute estimates 40 million people work at home and futurists forecast that number will continue to rise.

"Telecommuting is a trend that's starting slow," says Judith Louthan, a former senior help desk consultant for the AT&T/Global Information Solution Program Development Group (Louthan is currently Worldwide Help Desk and Business Planning Manger for Digital Equipment Corp.) "It's become accepted at level two [for second level analysts] but few help desks are encouraging level one agents."

She attributes this to limited switch capabilities, since only a few switch manufacturers (AT&T, Aspect and Rockwell are three) have built in the capability that allows agents to tie into the switch from at home and receive calls as if they were at the office.

But since switches need to be changed every 15 years or so, Louthan thinks more companies will choose new switches with this capability. "It's critical for first level agents to tie into the switch and access the knowledge

Telecommuting Impressions

A survey of 10,000 employees (from AT&T Global Information Solutions) found that:

- *Two-thirds cite better balance and increased productivity.*
- *Only 3% say it would interfere with family/home life.*
- *Two out of three say reduced expenses are a major benefit.*
- *92% are in favor of co-workers telecommuting.*
- *98% say their job category will include telecommuting.*

base when they work from home," says Louthan. "I encourage help desks to choose the right equipment to allow agents to telecommute even if they won't be ready for telecommuting for a couple of years."

In a traditional help desk it's common for first level agents to be able to answer 80% of all calls without any second level escalation when using a knowledge base. That leaves only 20% of calls that require second level support. When second level technicians are notified of an open call that needs attention via an e-mail message or through a pager they really don't need to be chained to a computer at the help desk. That's why it's more common to see second level technicians work from home.

Also, Louthan says that most of the questions second level technicians need to answer won't have to be looked up in a knowledge base, but are instead set in their heads.

Thomas Cross, chairman of Cross International (Boulder, CO) and developer of Nettalk LAN communication software, thinks that telecommuting makes sense for second and third level support staff. He says there's no need for them to be at the office since they can just be paged.

"They may need to do some investigating before calling someone back," he says. "They may need to go to a lab or check with others."

He attributes the increase in workers who telecommute to a change in workforce behavior. "People work under more relaxed conditions today," he says. "I call telecommuting the quiet revolution, because it's changing dramatically but quietly. Managers are just letting agents work from home, but the customer never knows it."

Time Management

"The ambience of the office is disruptive," says Cross. "When people work from home they can concentrate more. There are fewer distractions. People spend up to three hours a day commuting. Employees can complete the work for their entire workday at home in the amount of time it takes to commute. That leaves them with hours to figure out what to do."

Cross predicts that when these telecommuters run out of work they will call

How Telecommuting Saves You Money

LESS TURNOVER — which means you'll have long lasting workers and save on training costs for new hires.

CHEAPER RENT — You'll require less space and you can have three to five employees who telecommute share one cubicle for days when they come to the office.

SAVE ON PAYING FOR SICK DAYS — employees are more likely to continue to work when they are sick if they do not have to commute to the office.

the office for more work. "Then managers won't be able to accommodate them with more work and will have a major crisis," he says. He says that more work has to be designated up front for the telecommuters and productivity will soar as a result.

Who Should Telecommute

While telecommuting has many benefits for both employees and employers, there are some rough spots. Both Cross and Louthan agree there's some isolation involved when telecommuters are shut off from the rest of the office. There are also questions of fairness: which reps should you let work at home? And how can you monitor at-home reps?

Cross believes there must be a trust factor when allowing employees to work from home. "Managers need to look at employees and if they can't trust them they should fire them now," he says. "If you get voice mail when calling your employee and assume that worker is not a their desk because they are not working hard, that will carry over to the at-home worker."

Louthan says there needs to be a clear matrix for monitoring agents. "You need to have the technology to measure performance in place," she says. If your ACD lets at-home reps tie in as if they were at the office, then you should also be able to view the same statistics (number of calls, average length, logged in or out, etc.) as you do for in-house reps.

"If someone is not organized at the office, they won't work well at-home," says Louthan. She suggests that help desks put together guidelines as to what qualifies someone to work at home. "If a person wants to work at home they should be told what steps they must take before they can," says Louthan. "It's not an arbitrary decision."

Louthan cites The Travelers Insurance Group as an organization that has been highly successful letting agents telecommute. She says agents there can be compared to first level help desk reps working from home. Travelers uses telecommuting as a recruitment tool, allowing agents to telecommute after one year. They save $11,600 per telecommuter per year and have reduced turnover of claims agents by more than two-thirds.

Telecommuting's Future

Link Resources predicts the telecommuter population will grow at least 15% annually. Companies and workers will spend $3.7 billion equipping telecommuters in 1996.

By the year 2000 Louthan believes more companies will take advantage of the technology to equip at-home workers. "We'll see more switches that let agents log in from home," she says. "The cities that are part of the Clean Air Act will be ahead."

The Act says that companies from any of the areas the Environmental Protection Agency has singled out, must reduce the number of employees who drive to work between 6 a.m. and 10 a.m. by 20%.

Louthan says that just as ATM machines have sprung up instead of regional bank branches, it will be tough to find regional offices because telecommuting will eliminate the need to open regional offices.

Cross believes that the "revolution in telecommuting" will continue, but to overcome the possible isolation telecommuters may experience, neighborhood work centers will spring up. "People want to go someplace to work sometimes, like a park, but not everyday," says Cross.

"These centers could be a closed down school, for example, converted into a small office. Since workers will be more productive (as a result of telecommuting), companies will downsize, or at least be optimized and not have to hire more employees.

Ensuring Customer Satisfaction

It's a good idea to collect data so you can determine how customers perceive the effectiveness of the help desk. However, it's very difficult to measure exactly how customers feel. Not everyone will have time to answer survey questions (whether over the phone or through mail) and feedback from

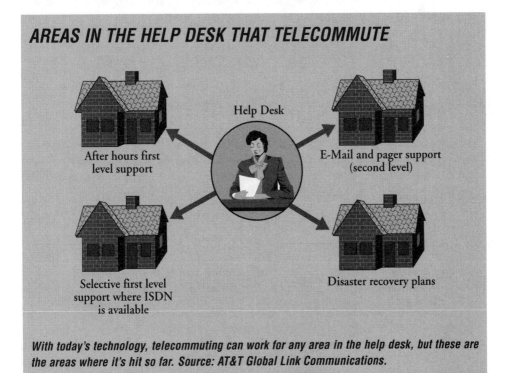

AREAS IN THE HELP DESK THAT TELECOMMUTE

Help Desk

After hours first level support

E-Mail and pager support (second level)

Selective first level support where ISDN is available

Disaster recovery plans

With today's technology, telecommuting can work for any area in the help desk, but these are the areas where it's hit so far. Source: AT&T Global Link Communications.

one person at a company may not be representative of how the company feels about service as a whole. Since people tend to remember most recent experiences as opposed to overall service, this may also skew results.

Still, requesting feedback from both customers who pay for support and internal employees is the only way to measure end-user satisfaction and get suggestions for changes or improvements. Customer information can help determine if the help desk is providing unnecessary services end-users do not make use of. Such information can also help determine if processes that end-users feel are inadequate should be improved. High customer satisfaction levels are rewarding to the help desk and give the help desk a positive image.

Ralph Tarantino says there are three effective ways to survey your customers: on the phone, person-to-person and through an official mailed survey. He thinks every help desk should survey their customers at least twice a year.

"You will get the greatest response if you offer a chance to win a gift [e.g., software or a PC]," says Tarantino. "A 10% to 15% response rate through mail is normal, but 30% is more likely if you offer a chance to win something if they return the survey."

How Hewlett-Packard Measures Customer Satisfaction

Customers who phone one of Hewlett-Packard's 24 toll-free numbers are typically greeted by an interactive voice response system. In 1989, HP used no automation. Now more than half of their calls go through an automated system, says HP's Steve Elias. To find out how their customers felt about the automation, the company's Santa Clara office asked customers to answer satisfaction surveys over the phone using IVR or by requesting a faxed questionnaire.

Fax-Max (Los Altos, CA) is a systems integrator and reseller for Ibex Technologies (Placerville, CA). They recently installed a system called FactsLine at HP's Santa Clara office.

Now after callers retrieve the information they need through the IVR system, FactsLine ask them if they would like to answer a survey by pressing digits on their phone pad or by receiving a fax. When they enter numbers for corresponding answers, the system captures this information in the database.

If they choose the fax option, the system sends the customer a fax and when they send it back, the system reads the choices they made and stores the info in the database.

A HP manager can then retrieve the data and import it into graphs or other formats. "This information allows us to give feedback to other departments with customer backing, to show them why they need to automate," says Elias. Elias says they had been using an outside service bureau to collect customer satisfaction data, but this system is more effective and cost efficient. "It's pennies to the dollar," says Elias.

"We have so much versatility to solicit data and automatically tabulate it. We can turn (the system) on and off at will. We get powerful data from these surveys and our customers appreciate the minimal effort to complete their transactions."

He says another effective means of surveying is to do it randomly, calling back the twentieth or fiftieth caller every day. "You get their feedback while the call is still fresh in their heads," he says.

Hiring an outside agency to put together surveys or poll customers is a good idea for larger help desks that want to ensure impartiality and professionalism.

Tarantino recommends asking what bothers the customer most and what they would like to see improved; asking if their problem was handled in a timely manner; how they perceive the skill level of the technician; and whether or not he or she was courteous.

Many customers may name the employees who always do a great job helping them or who they hope they never get when they call. Such surveys can assist in identifying training and staffing needs.

Sample Customer Satisfaction Survey

Here's an example of what a typical customer satisfaction survey should look like:

Please answer the following questions by checking the column that most closely fits your feeling.

	excellent	very good	fair	needs improvement	unsatisfactory
Support rep's helpfulness					
Support reps's politeness					
Ability to explain the problem					
Amount of time to solve problem					
How did he or she meet your exceptions					
Speed of answer					
Service hours					

1. In a typical month, how many times do you call the help desk? _____

2. What products do you most often require support for? _____

3. What service are you most happy with? _____

4. Least happy with? _____

5. Give us two suggestions for improvement, if any. _____

➲ Chapter 8:
HIGH-PROFILE CASE STUDIES

ABC'S RESPONSE CENTER

ABC's response center in Manhattan supports 4,000 end-users from all over the country. The response center receives about 1,200 calls a month — mostly regarding mainframe (they use mainframes at their Hackensack, NJ office), local area network and PC problems.

"The amount of calls we get has been climbing because Capital Cities has been investing in information systems and training users," says ABC's Keith O'Sullivan. He says that as more people within the company become PC-literate the more the response center will grow. "We're a Microsoft shop so there are application problems too," says O'Sullivan.

They also use a wide area network. "Because of the nature of our business we have correspondents (news reporters) who travel around the world and need to dial in remotely to the center." Their server at their London office can talk to their Manhattan server, he says.

They have four first-level analysts who are trained to answer PC, networking and mainframe questions. Their second level analysts are hardware and software specialists. They solve about 60% of their problems at the first level. "We're trying to get up to 75%," says O'Sullivan.

For problem resolution they use Support Magic from Magic Solutions (Mahwah, NJ). "We're using the system to build a database," says O'Sullivan. "We can pull up a caller's history to see if the problem is chronic. The system lets us search on anything." He says it's helpful to be able to call up things like the number of past problems on Microsoft Word.

Employees on their mainframe use the Magic system too, but sometimes there's a delay. "We're going to be upgrading to Magic's SQL version which will eliminate delays encountered when different users go across the wide area network," says O'Sullivan. "It will speed up the response."

They are also in the process of installing a Nortel Meridian ACD, and are planning to install software so analysts can remotely take over end-user desktops to make problem solving easier.

BAY NETWORKS

Last year, Synoptics, manufacturers of networking devices, merged with Wellfleet Communications (router manufacturers) to form Bay Networks. The new company manufactures and distributes high tech networking equipment.

Currently they have a major support center in France and two smaller centers in Australia and Tokyo. In the US they've formed Bay Networks West and Bay Networks East. "We're hoping to integrate our Aspect (San Jose, CA) and Clarify (San Jose, CA) help desk systems to create a global virtual call center," says Bay Network's Rusty Walther. In such a center their systems would look at the time and type of call coming into the center and route it to the best equipped center to handle it.

For now, they have formed a neatly integrated system at their West Coast center. This is how it works: When a call comes in, their Aspect ACD prompts contracted customers to enter their site identification number.

The ACD sends the entered number to the Clarify (San Jose, CA) Clear Support system, where the system looks at the customer's contract and checks its validity before sending the contract status and priority level back to the Aspect system. The customer then returns to the main menu to key in their type of problem. The customer is connected to an engineer who knows the category of their problem and is trained to solve it.

Customers who do not purchase a maintenance contact for phone support can choose to pay $150 per incident. "Whether it takes one or several calls, we'll solve the problem," says Walther. Since offering a pay-per-incident support system, they've sold more regular service contracts. "Pay per incident gives customers an understanding of the value of a support contract," says Walther.

When callers opt for the pay-per-incident support, the Aspect system

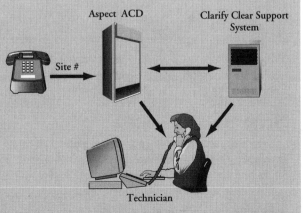

Contracted Customer Support

When a call comes in to Bay Network's support center, their Aspect ACD prompts contracted customers to enter their site identification number. The ACD sends the number to the Clarify Clear Support system, which looks at the customer's contract and checks its validity before sending the contract status and priority level back to the Aspect system. The customer then returns to the main menu, pressing a phone key to indicate his or her type of problem. The customer is then connected to an engineer who knows about their problem and is trained to solve it.

Aspect ACD

Clarify Clear Support System

Site #

Technician

prompts them to enter their credit card number. The system transfers this number to the Edify (Santa Clara, CA) Electronic Workforce system. A "software agent", part of the Edify system, verifies the credit card number, then sends a contract to the Clarify system so it can update the database.

The Edify system faxes the customer a receipt. Once this is complete the Aspect system announces to the caller that they have verified their credit card, tells the customer what his or her site number is, and sends the caller back to the main menu.

The call is then routed to a specific engineer group who can solve the problem. The call receives the same priority as a customer on a regular support contract. Another voice prompt gives the caller the option to go into their IntelliSystems (Reno, NV) voice response system to try and troubleshoot their problem without the aid of a support rep. Lower priority calls come from their technical partners, resellers and field technicians.

They use this tightly-integrated automated set-up to front-end an average of 800 calls each day. These are handled by 45 engineers from 6 am to 5 pm Pacific time. Their east coast office in Billerica, MA, stays open 24 hours a day.

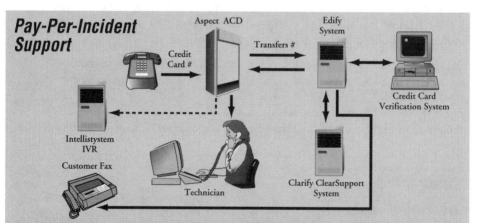

Pay-Per-Incident Support

Customers who do not purchase a maintenance contact for phone support can choose to pay $150 per incident. The Aspect system prompts callers to enter their credit card number. The system then transfers this number to the Edify Electronic Workforce system. A "software agent" (part of the Edify system) verifies the credit card number, then sends a contract to the Clarify system so it can update the database.

The Edify system faxes the customer a receipt. Once this is complete the Aspect system announces to the caller that they have verified their credit card, tells the customer what his or her site number is, and sends the caller back to the main menu.

The call is then routed to a specific engineer group who can solve the problem. The call receives the same priority as a customer on a regular support contract. Another voice prompt gives the caller the option to go into their IntelliSystems IVR system to try and troubleshoot their problem without the aid of a support rep.

"(This system) saves a huge amount of time," says Walther. "We only need to have one or two people front-ending the switch for callers who have rotary dial or get lost in the prompts."

He says that without such a set-up they would need ten to twelve people fronting the switch. He says it also saves a huge amount of time — their engineers know about the customer and his or her problem when the call reaches them.

KRAFT FOODS

When you think of Kraft Foods, the first thing that probably comes to mind is cheese. You probably don't wonder how they support their close to 4,000 employees when you see their products on supermarket shelves.

But at their White Plains, NY support center they support all of these employees plus about another 1,000 distributors at brokerage houses who sell their products to supermarkets.

"We have account executives in the field who use our communications software to generate invoices and get volume information," says Kraft's Mike Brown.

The problems these salespeople may run into in the field include malfunctioning modems, laptops that run out of memory or that can't connect to the mainframe when dialing Kraft's toll-free number, says Brown.

To help solve such problems, Kraft uses software from Professional Help Desk (PHD, Stamford, CT). "We can type in an error number or key words when describing the problem to find a possible solution based on how we solved a similar problem in the past," says Brown. They built their own knowledge base and solve between 70% and 80% of problems from the knowledge base.

In the past the process was all manual. They would write things down on pieces of paper. "Now we can find answers a lot quicker," says Brown. "We spend less time on each call." He says they also use the system to get information about hold times and the number of open calls. If a call has been open too long the system forwards it to an appropriate technician.

MERCEDES BENZ

The help desk of Mercedes Benz gets about 25,000 calls a month from dealers throughout North America and internal Mercedes Benz employees. Their dealers work on PCs and use software from Mercedes that tracks sales information, inventory and other automotive related information. At their Montvale, NJ help desk Mercedes has three high level analysts who field dealer calls and four support reps who handle employee calls. Employee calls can come from any of their six regional offices or from their 150 field engineers who work on IBM Thinkpads.

"All of our analysts are cross-trained," says Mercedes' Ellen Richardson.

"They're able to back each other up." They also have a second level group of programmers who handle problems that get escalated to them. "We had been using a call tracking system on DEC mainframes, but with the industry changing, we wanted to phase out our mainframe system and start using Windows," says Richardson. "So we needed client/server-based help desk software."

They recently started using Heat for Windows from Bendata (Colorado Springs, CO). "We liked the product's ease-of-use," says Richardson.

"The system helps us solve about 75% of problems on the first level. We use decision trees to find solutions in our knowledge base." She says they also purchased some KnowledgePaks from ServiceWare. "The system's call journal helps us see what's going on and what problems each analysts is working on."

MTV

When viewers watch MTV, VH1, Nickelodeon or Nick at Night, they're not thinking about the behind-the-scenes people necessary to make their viewing possible. MTV makes sure someone is available internally 24 hours a day to ensure things run smoothly.

All four of these cable channels are part of MTV Networks. Their internal support center is designed to serve critical needs. When 5 p.m. rolls around programming does not stop. That means technicians must be on call for any user problems.

During working hours MTV Networks has 10 to 15 technicians on hand to answer any questions or help with technical problems. "People call with questions about anything — from how to access the computer network to wanting to know a forgotten password," says MTV Networks Jaimes McNeal.

Their Manhattan office takes calls to help support personnel in their Los Angeles, New York and Chicago offices and satellite locations. Each office shares the same database, running across a wide area network.

These technicians rotate phone time so that at any one time four to six technicians are logged in to take incoming calls. When off the phone, technicians keep busy getting dispatched within the building to deal with situations face to face. Calls come in on a Nortel Meridian ACD. During off-hours the ACD goes into night mode and calls are routed to an answering service. They get about 600 help desk calls each week.

"The answering service acts as a filtering device," says McNeal. "Analysts have computers at home set up the same as at the office. They have two phone lines so the answering service can call them. If there is no answer the service bureau will page them."

As a back-up resource they will also try to contact another analyst. If that too fails, a third level of support connects the caller with an IBM help desk, familiar with MTV Networks' support center and able to answer how-to questions. The next working day the answering service sends MTV a fax of all activities.

To help answer questions and track calls, they use C-Tam (Customized Telecommunications Asset Manager) from Walsh-Lowe & Associates (Hoboken, NJ). The system tracks their TV equipment all over the world. An inventory module keeps track of the equipment and serial numbers as equipment gets moved from one place to another.

CTAM's Help Desk Management module lets technicians get inventory data, equipment types, common problems and recommended solutions. The system is customizable and technicians only need to point and click to get the information they need. Trouble history is accessible by equipment type, location or user.

NOVELL

Superior technical support from one of the largest networking vendors in the world is a complex but efficient process. At their main headquarters in Provo, UT they get about 130,000 calls a month.

These calls include requests for product information, major account customers who need support at stage one and calls from their channels (distributors, resellers and other partners) needing advanced support for questions or problems they can't solve locally.

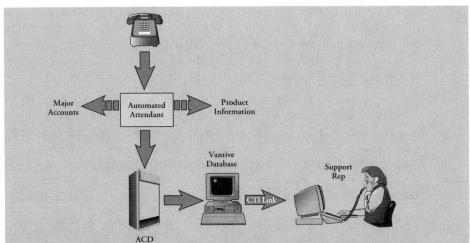

If the customer calls Novell back after reporting a problem, an automated attendant asks him or her to enter the incident number for the problem. A CTI link between Novell's Aspect ACD and the Vantive system redirects the caller back to the owner of the problem. This rep receives a "screen pop" full of customer information with the call.

Their channel partners provide on-site and front line support for Novell customers. The Novell help desk serves as an escalation point.

The company had been using a home-grown "basic call tracking system." But feedback from their customers indicated their strategy should be to move towards a more technical support solution. Their home-grown system was not at a level where they could integrate with their phone system and re-use it.

They chose Vantive's (Santa Clara, CA) Support and Tools modules for customization, which allows them to more easily fulfill their strategy. Novell offers service around the clock. For example, if it is after 6 p.m. Mountain Standard Time, calls get directed to another center where the agent knows who the caller is and has all of their historical information.

The customer support rep looks at the caller's record and asks for a brief description of the problem. Then technical support engineers search the database for a solution. If they find no solution or the problem requires further research, progress towards resolving the incident is captured in the database.

If the customer should call back after reporting their problem, they are asked to enter the incident number. A CTI link between Novell's Aspect (San Jose, CA) ACD and the Vantive system redirects the caller back to the owner of the problem. The engineer gets the call and a screen pop.

Off-line, Novell researches the problem so when the customer calls back they have answers and recommended solutions to make them more efficient. This also helps to eliminate "phone tag". They have received positive feedback from customers who've called back.

An after-hours US call could be routed to one of their other primary centers in Dusseldorf, Germany or Sydney, Australia. A recording informs callers of the overseas transfer. They pre-announce that the call is being transferred because they found that callers like to be notified when they're being transferred to another country.

The largest percentage of their calls are questions about how to bring multiple components together to build a total solution. Their goal and strategy is to add value to multi-vendor solutions and eliminate finger pointing, working as partners. This way, they can offer greater value to customers and customers can just call one number for complete support.

QUANTUM HARD DRIVES

Quantum is the world's largest supplier of hard disk drives. Their success is astounding — they employ 9,000 people worldwide and sales for fiscal year 1995 rang in at $3.4 billion.

Their customer base is comprised of PC manufacturers like IBM, Dell, Apple, HP, Compaq and DEC as well as general consumers. They break down their support center into two parts. One is Distribution Support, which handles 25,000 mostly simple calls each month from consumers who call in on a single toll-free number.

On the other side is OEM support. Here they get fewer calls, but the calls are much more complex and critical since their OEM business represents about 70% of their annual revenues. The issues their highly skilled engineers deal with typically concern prototypes of new PC products using Quantum's storage technology. These issues take weeks or even months to resolve because they involve new products in the early stages of development.

"Here we have account teams dedicated to each OEM's account," says Quantum's Fred Osowick. "We have field offices worldwide located near OEM sites so calls that can't be resolved are escalated to OEM field engineers who can go to the customer's site."

Osowick is Quantum's technical support manager for distribution support. He says they have 18 support reps who use ClearSupport, help desk software from Clarify (San Jose, CA) and Clarify's Full-Text Search Diagnosis Engine.

Calls come in on a Nortel (Richardson, TX) ACD and are routed to support reps. But if there are long hold times, callers are greeted by Edify's (Santa Clara) Electronic Workforce IVR and fax-back system.

"If there are long waits in queue we like to give callers a choice to use automation," says Osowick. Prompts are typically something like this: "press 1 to request a manual, press 2 to reach our bulletin board service, press 3 to go into our fax-on-demand system and press 4 for live support."

Osowick says automating has made for a great return on investment. Just three years ago they were only handling 6,000 calls a month but had 24 technicians. Today they can handle four times the amount of calls with a third less staff.

To solve problems, Osowick says they use the system's Diagnostic Engine to search the knowledge base. He says there are four scenarios to bring a drive to life (a typical problem). To search the knowledge base, reps use key words

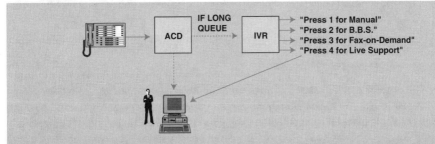

When callers dial into Quantum's distribution support center they can be routed to a support rep, but if hold times are long, they are instead routed to an IVR system. Here callers can make selections that may give them the information they need and eliminate the need to speak with a support rep.

and true and false options. "If the system comes back with two suggestions and one is linked to 10 cases and the other is linked to 1,000, we'll suggest the customer try the solution linked to 1,000 other cases," says Osowick.

Engineers in OEM support also use the Clarify system. Quantum has about 120 concurrent ClearSupport licenses. "We're looking to tie the system into everything we do," says Osowick. "When you can't spend the money to hire more people, you need to automate."

STAPLES

Visit any Staples store and you will be overwhelmed. This super store chain carries everything a person could every need for the office. Up until a couple of years ago they had only forty stores and no main support center. Today they have more than 350 stores throughout the US and Canada.

All of their stores have an AS/400 terminal connected on a Wide Area Network. This setup ensures that if a customer wants a particular item not

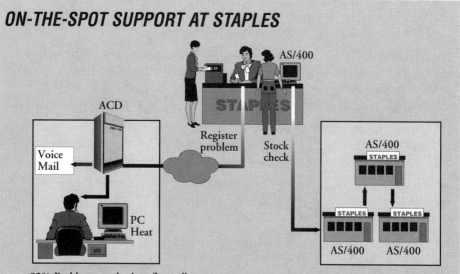

ON-THE-SPOT SUPPORT AT STAPLES

82% Problems resolved on first call
Average resolution time >3 minutes.

If there is a register problem at any Staples store, an employee at the store calls their main support center to find out how to correct the register problem. Speedy answers are a priority since the cashier is often in the middle of ringing up a customer who is waiting at the register. The call goes through their AT&T ACD and is passed on to a support rep who can search the knowledge base using Bendata's Heat to try to solve the problem.

In a separate incident, when customers ask for a particular item not in stock at the store they are in, a store manager can go to their AS/400 terminal connected across a WAN to terminals at their other stores to see if one of the other stores has the item in stock.

found in the store they are in, a Staples employee can check to see what other stores have the item.

They use Heat from Bendata (Colorado Springs, CO) to track calls and help solve problems. Staples' Eric Helbing says they are in the process of upgrading to Heat for Windows from the DOS version.

"We operate on a Windows NT network and use a suite of products from Microsoft," says Helbing. "Without Windows we're not getting all the power from Heat that we should be." For example, he says sending e-mail will be much less cumbersome once they are in Windows.

Over the past two years they've built a comprehensive knowledge base that helps them solve about 82% of their problems on the first call. This is critical since about 80% of their calls come from stores that are having a register problem, often while ringing up the customer.

"We have four or five agents taking calls at any given time and our queue time averages less than three minutes," says Helbing. Callers with less pressing matters have the option to leave a voice mail message.

Until a couple of weeks ago, when they installed an AT&T ACD, they were using Heat to track the number of calls they received. "It was the only way we knew how to staff," says Helbing. "Our goal is to have all of our users on Heat and go completely paperless." With their new site license purchase, they're about to roll Heat over to their second level technicians. "We used to print tickets and hand deliver them to second level," he says. "Now these users can log on to check open tickets. We can automatically notify them."

Smooth Sailing Tips

• Get organized •Know what resources you have and assign them appropriately • Keep your support staff trained and equipped with current systems • Wherever viable, integrate with other products for faster database population and dealing with technical management issues. • Build usable problem escalation procedures. • Build usable change management procedures and include those with sign-off authority. • Select a consistent & organized naming scheme to identify clients, accounts, devices, sites and location codes.* • Assign responsibility for the database system (coordinating product & resource changes, distribution on library files, etc.).** • Assign responsibility for billing accounts (to post time & materials to budgets, manage billing and service contracts etc.).** • Track your calls routinely to determine problem patterns and help create pro-active ways to reduce calls. • Stay in touch with your help desk vendor and let them know what changes you could use.

(* If this scheme allows for easy recognition then your support staff and network users will all find it simple to use.

**This typically sounds more overwhelming than it is but particularly on larger multi-user systems your system's functionality will soon fall behind if no one manages this.)

— STEPS: Tools Software, A Division of Cauchi Dennison & Associates Inc.

➲ Chapter 9:
THE FUTURE OF SUPPORT

The way help desks operate in ten years is certain to be a lot more efficient. We're already seeing the way end-users seek support changing. Support costs will drop for both the help desk and customer as more end-users begin to realize that using self-support technologies to solve their own problems is less costly and time consuming than calling the help desk.

The technologies will improve, leading to more widespread and effective use of the Internet, intelligent interactive voice response systems and techniques like fuzzy logic and natural language processing for problem resolution.

Opinions

We asked a number of help desk vendors what they envision to change in the future of support. Here's what they said:

The future of support is going to be extremely strong! Support over the Internet will be huge as it provides low cost, reliable and fast access to locations anywhere in the world. Voice communications bundled with support packages that provide remote access and remote control should be very popular and in demand. Support requires products that are easy to use, fast and reliable. Products that save companies money will be imperative to the support market.

We will see more embedded use of tools for problem identification and resolution. For example, expert systems tools like case based reasoning lend themselves easily to "agent" operation, and can be integrated with desktop systems for low level information streams. This will enable rapid diagnosis of problems as they occur and in increasing levels, automatic problem resolution of the most common occurrences on the desktop. As the sophistication of these embedded problem resolution tools improves, we can expect to see support centers connected directly to user systems with the ability to interact with those systems to resolve problems in a proactive manner.

— *Avalan Technology*

Fully automated assistance without individuals. With the IBS technology we are headed in this direction. Using an SOS icon, information could be passed from the user's workstation to a central help desk database. The database could pass specific data relating to the user's solution back to the workstation or even resolve the problem with permission by executing a recorded script.

— IBS Corp.

Support is a much more global issue than simply being the domain of the traditional back-room system support group or the local LAN administrator. We see enterprise management as something that will ultimately come to mean change management of all corporate resources NOT only hardware, software and network connections but also staff and outsourcing capabilities. This means that the network professional needs to know what and who is out there as well as having a consistent way in which to manage events. This is important as businesses streamline their resources and continue to move their line operations, along with the management and tracking of them, into the systems arena.

We believe that more organizations are currently refocusing their efforts to find a consistent and deliverable approach to overall change management that includes the capabilities found in available high tech systems. We see this as a resurgence of more centralized control. However, unlike their mainframe predecessors, these systems will be more flexible, scalable and available. For this reason, we approached the help desk issue from the project management/change management perspective and delivered a scalable product that functions in both a standalone and integrated fashion. This allows users at all stages to solve their immediate problem while moving toward a more global solution.

— STEPS:
Tools Software, A Division of Cauchi Dennison & Associates Inc.

The industry is going toward more self support, and willing to use more sophisticated self help tools, including the Internet. There will be more interconnectivity between vendors and customers; thus more electric communications as supposed to people using the phone. Increasing acceptance of telephony. Non-support people will be able to log on calls & search for answers themselves.

— Ron Muns,
VP of Business Development, Astea International Inc.

We are beginning to see the mode of moving support to the point where support is needed. This can be accomplished in two ways: for the problem to self-correct or for the person who is having the problem to solve it independent of the call center. Recent changes to the market, including access to the web and problem resolution software, all point to pushing support to the end-user.

— Eileen Weinstein,
VP Operations, Advantage kbs

The integration of client/server enterprises poses an interesting challenge for help desk vendors, as this is the area most frequently tapped by companies as the center for the enterprise. Due to its control over vital information about customers, system configuration, change management, asset management, product bugs and more - the help desk is a logical place through which mission-critical applications should flow — and this is increasingly the case in Fortune 1000 and higher companies.

As the help desk becomes the hub for client/server enterprises it follows that these applications must not only be flexible enough to fit the organization's unique needs, but must also have the adaptability to change as companies needs and processes change. It also becomes apparent that the help desk application being used must be able to fit into the most heterogeneous of environments — fitting with any database, any server, any client platform as well as integrate with other key technologies such as telephony, beeper notification, network and system management and more.

— Remedy

The Network Management group and the help desk will be reorganized under the IS structure to be more integrated. The roles and responsibilities of each are starting to overlap and their customer base is the same. The software tools designed to support these functions will also follow the same path of integration.

— McAfee

The help desk will undergo almost revolutionary change in the next three to five years, with existing help desk software and practices destined for total overhaul. These factors will drive this change:

As PCs become better integrated with enterprise computing, organizations will no longer tolerate the chaos caused by differences in PC configuration.

The "network terminal" concept will triumph but not as an under-powered Web terminal — rather as a powerful personal workstation that includes a complete suite of office productivity software.

Individual users will access server-based applications that download office-compatible components to the workstation's large disk drive. Replacing unique PC configurations by a standard PC setup that just happens to have a very large "personal cache" will eliminate many of today's computer problems.

Tomorrow's standardized desktop will be similar across many companies and therefore more cost-effectively supported by third-party "outsourcing" vendors who will have the expertise and capacity to set up high-volume, high-quality operations (probably using off-shore personnel).

This means that the internal help desk will be able to turn its attention away from petty technical issues towards the real support crisis: helping organizations get business value from their computer systems. Expect to see the typical "Help Desk 2000" helping users achieve business goals with internal applications that embody competitive advantage while an outsourcer helps them reset their password.

— PLATINUM Technology

Asset Management Today's leading help desk vendors recognize the need for a support center solution that goes beyond simple call tracking. Asset management has become an essential component to a comprehensive help desk solution. Help desk software should facilitate complete asset management of an organization's hardware and software. Maintaining inventories, analyzing depreciation, tracking warranty and service contracts, planning upgrades and ensuring license compliance help to maximize return of investment.

Help desk software must go beyond simple call tracking in order to successfully manage every aspect of today's support center operations — it must provide flexible, powerful call handling and tracking features, along with comprehensive asset management capabilities.

Service Level Agreements As Information Technology grows to be more complex and distributed, it be comes increasingly difficult for support organizations to solve problems on the first call to the help desk. Today's busy support centers are using the sophisticated practice of Service Level Agreements (SLAs) to successfully handle the multiplex of calls that come through the help desk. SLAs ensure that all calls are managed in true business priority order, and that the support center is providing a level of business continuity that meets its customers' expectations.

Recognizing the importance of this growing problem, Datawatch

Corporation has made SLAs central to the workflow of Q-Support, its Windows-based help desk and asset management software.

SLAs offer a more sophisticated approach than simply setting priority cod codes to a call. In most cases, priority codes are insufficient because they fail to reflect the specific needs of different areas/groups within an organization, and they don't show how long a call has been in the system. A low priority call may wait indefinitely while higher priority calls are resolved. And if the calls are raised in priority, then all calls could be raised to the same level, leaving no method for determining which calls to address next. SLAs provide the complete solution.

Service Level Targets (SLTs) are set within each SLA. The SLT specifies a period of time that the support center is expected to respond to and fix a certain type of call. This time frame can vary according to the area within the organization (department, group, etc.), the type of problem, the equipment, or the services affected. For example, a typical agreement might state that a call must be responded to within 10-minutes and fixed within two hours, with service provided between 9:00am and 5:00 pm Monday through Friday.

SLAs provide support centers with an effective means to maintaining quality service standards. Whether it's service from your help desk or from third-party suppliers, tracking compliance with established standards and contractual obligations can be made easy with SLAs. SLAs set customer expectations so that the support center and its customers are aware of the time frame and parameters in which a problems should be resolved. And by measuring service level performances, support center managers can easily determine if the support center is performing optimally.

— Datawatch

Customer support requirements continue to expand as the complexity of software products increase and end-user expectations grow. At the same time, many support organizations are finding themselves under tighter budgets. To meet this demand, they have turned to call management and problem resolution systems. While helping to address some of the issues, these systems are many times incomplete since they do not contain the actual information used to solve a customer's problem. Embedded knowledge bases — ones that run native in the problem resolution system — are the answer.

By using embedded knowledge bases, customer support organizations are able to help end-users resolve problems more quickly and less expensively. In addition, they reduce the time required to solve a problem, and they significantly reduce the number of calls that are unresolved at the first-level help desk. Finally, they improve the quality of support, provide more consistency in help desk answers, and create an ability to deliver knowledge directly to end-users. All of these benefits help the customer support organization gain the full value of their problem resolution tools.

— ServiceWare

Customer support will increasingly be a key company/product differentiator and demand for support will continue to increase. Support tools, such as IVR (and speech recognition at some point), enhanced PC help systems and knowledge base searching via the WWW will increasingly push self-help out to the customer. Advanced problem resolution tools will finally become easy to implement and use.

— Opis

Two key technology trends have influenced the industry in the last five years: the use of knowledge-based systems for problem solving and the inception of the Internet, allowing call deflection or call avoidance in call centers.

1. The use of knowledge-based systems for content navigation for problem identification and resolution which were traditionally deemed too difficult for standard commercial use, took a quantum leap forward in the 1990 s with the advent of the commercialization and proof of case-based reasoning (CBR) systems which for the first time allowed for the broad use of this type of technology. CBR is now the defacto standard expert systems tool in the call center market. Eight of the top ten 10 help desk tool providers use CBR.

2. The Internet is enabling direct customer access to knowledge bases maintained by call centers. This is allowing the implementation of call deflection/call avoidance strategies where call centers can offer "self service" for problem solving, product selection and general information.

— Inference

Just as the '80s were the years of the Quality Initiative, Software Artistry believes that successful organizations in the '90s are committed to the Support Initiative. Today's corporations are faced with the challenge of improving their operational efficiency while implementing increasingly complex technologies. At the same time, they are competing more intensely on a global scale and striving to provide superior customer service as a competitive advantage.

To meet these multiple, interrelated challenges, the support function within an organization has become the foundation of the enterprise. And the traditional help desk has evolved into a next-generation strategic support center that is tightly linked with other functional areas.

Recognizing the inter-relationships between these separate areas within an organization's IT framework, Software Artistry's EXPERTISE provides the Enterprise Support Management tools required to link operations as diverse as the help desk, network and system management, change and asset management, and end-user empowerment. This last area is another important trend in the support industry - the direct empowerment of end-users and customers so that users may report a problem, incident, or request information without involving a support analyst. The "call" is then responded to just as if it came in over the phone. Software Artistry is committed to this "forward deployment" of information through its products Expert Web, Expert Mail Agent, and Expert Access, all of which help reduce call volumes in support centers.

— Software Artistry

Providing competent, available technical support continues to be a critical success factor in all technology businesses. As technology vendors and end-user organizations employ technical support solutions to gain competitive or productivity advantages, we expect to see these changes occur in the support delivery landscape:

1. The development of more web-centric support tools - most technical support tools available today are LAN and database-centric solutions. While many have an interface to the WWW, the tool itself still runs in a LAN environment and is limited to running on a limited set of supported clients. Our vision for the future of support automation is to deliver a web-centric tool that could be delivered, for example, as a series of Java applets to the client's desktop. These applets will perform these functions:

- provide the interface for the support agent to interact with the support system from any web browser anywhere on the Internet or an intranet.

- provide the means for the support customer to obtain support electronically.

2. Tighter integration of support systems with other applications that maintain customer data - currently, most help desk and technical support solutions are not integrated with sales tools, contact managers, software metering tools, equipment/inventory management, parts ordering systems or accounting systems. We believe that this will change and most or all tools used to manage any type of customer transaction will be consolidated into a single repository of customer data accessible by a series of commercially available integrated applications. While solutions like this

do exist today, we expect to see even tighter integration between systems and greater affordability, opening the market up to more medium and small-sized companies.

3. The use of collaborative technology to improve the problem solving process and speed communications between customers, support agents, and product engineers.

— Teubner & Associates

In the immediate future, customer support centers will come under increasing pressure to provide excellent service for less cost, leading them to apply innovative technologies and approaches to fulfill their mission. Technologies and business models that allow them to solve customers' problems faster, allow customers to solve their own problems, and also convert the support center from a cost center to a profit center will all become mainstays of the most advanced support organizations.

— Primus

The future of support will develop in several ways. Very highly centralized, large corporations will attempt to tie in the internal help desk with customer service and other aspects of the organization into one large system. Moving forward, however, we feel that the needs of the internal support and the external support systems will be increasingly different.

In the internal support area, where corporations are supporting their own employees, more unified solutions that incorporate network management, LAN management, support, training and change management will all be more closely tied together. These solutions will primarily be of interest to decentralized operations that have multiple help desks and IS functions that may want to share information with one another. Specialists in the various areas will cross-integrate and cross-sell one another's systems, so that a best-of-breed solution is delivered without compromising how the applications can work together.

We think that internal support solutions for the help desk area are pretty mature and handle the needs of the help desk well. The challenge will be integrating that technology with related support areas, and making the most of the distribution channel that will get you not only in front of the help desk manager, but also in front of the person that handles asset management issues, etc., since you may have other integrated products that fulfill those needs.

Problem resolution will continue to develop so that self-help options will be more viable for the end-users (we don't think they are now) and knowledge bases that handle commercial software troubleshooting will become more common, so organizations can concentrate on developing answers to the problems that are specific to their organization, not a lot of time on Word problems, etc.

The needs and requirements of the international community will continue to drive the industry in some different directions, since other cultures will have slightly different needs and areas of interest in products, and the vendors will have to move to meet those needs.

— Utopia Technology Partners

In an increasingly-fragmented and hyper-competitive marketplace in which the cost and complexity of information systems and customer support are escalating rapidly, help desks and call centers are quickly being transformed from simple "trouble ticket" departments into vitally-important strategic operations for the modern business enterprise.

Scopus Technology sees three major trends emerging that will shape tomorrow's support efforts:

1. Integration of the Internet and World Wide Web with existing support systems will be crucial to providing anytime, anywhere support for both employees and customers alike. As Eric Schmidt, Chief Technology Officer for Sun Microsystems put it, "Customer service is the killer app of the Internet."

2. Personalization of support services will not only be critical to transforming customer problem reports into new sales opportunities, but will also increasingly provide companies with a crucial competitive advantage in the battle for market share and customer "mind share." If "mass customization" and "one-to-one marketing" are the pillars of future business success, then personalization is the key building block.

3. An Extended Enterprise approach to support — one that involves not only various departments within the company in the support process but its outside vendors and partners as well — will be absolutely vital to providing effective customer support in tomorrow's increasingly-outsourced and multi-vendored competitive environment. Without this Extended Enterprise approach, the customer feedback loop, so crucial to keeping tabs on changing customer needs and market conditions, will be broken.

— Scopus Technology

1. Increased external competition and internal management pressure will combine to demand better customer satisfaction (faster response times, wider variety of services, longer support hours, etc.) at a lower cost.

2. Continuing changes in communication (the Internet, corporate Intranets, LANs, WANs, etc.) will increase both the demand for and the ease of sharing information. Providing direct access to knowledge bases will enable customers to get answers to their questions and problems more quickly and economically.

3. Information stored in support organization databases will become increasingly valuable as companies strive to better understand their customers' needs and requirements. This knowledge will enable them to increase customer satisfaction, develop desirable enhancements to existing products and create additional products. The demand for complete and accurate information will increase as companies review the data collected by customer support from multiple perspectives (e.g.: sales, marketing, training and finance).

4. Continued demand for the elimination of defects to meet total quality management goals will require support departments to carefully track and demonstrate progress in this area. Corporate demand for better ROI (Return on Investment) will increase pressure on support departments as product development cycles are shortened.

— Repository Technologies

Future changes in support will see increased integration with other internal business units and an increasing emphasis on going beyond the basics of problem-solving toward establishing an ongoing, two-way relationship with the customer.

As technology enables ever more complex products, support providers will have to master more specialized tools. At the same time, customers will continue to demand outstanding service and support.

— ProAmerica Systems

With organizations fervently working to optimize operational efficiency and streamline as many processes as possible, the effect on support can probably be summarized in two words: distribution & consolidation. Distribution of information, that which can help both users and customers solve their own problems, is today only starting to be realized with the accessibility of information over the World Wide Web. This trend of "recipient-initiated support" will continue in not just the use of distributed access to problem resolution tools, but also in terms of more robust distributed diagnostic applications running on clients' desktops.

Consolidation will also be a major theme in the future of support, as more and more organizations realize that "user support" or "customer support" is merely a narrow niche in the more broad spectrum of information management. Organizations will look to consolidated technologies (today they are called "Enterprise-wide applications") that address all aspects of information tracking, processing, and retrieval, from facilities management to work force automation.

— Applix

Support organizations must become more customer focused. There must be more options for service to suit the varying personalities of the customers (i.e. direct in person support, support technology accessible to the end-users deployed to either a local server, an Intranet or the Internet, or directly onto the end-user desktop). Support organizations must become more proactive and must be able to translate their value into a direct dollars and cents impact on their customer's bottom lines. Technology may not always be the answer to support problems, but the right technology, chosen wisely, can be a great benefit.

— Amdahl

While resolving incidents logged by customers is a key role of a help desk or support center, the true benefit is derived when overall corporate productivity and efficiency is improved. To accomplish this, the ability to resolve incidents must be pushed as close to the customer as possible. Empowering the person receiving the initial customer call with the knowledge to resolve incidents is critical.

These first level individuals have the ability not only to resolve incidents, but to record their occurrence thereby providing critical data for the future product enhancements or purchase decisions.

While the technology exists today for customers to research and resolve their own problems, the feedback loop is incomplete in that a database entry is not made to indicate that an incident has been resolved.

Support in the future will further drive technology to allow customer problem resolution while completing the feedback loop.

— Silvon Software

1. As technology lowers the cost of entering any playing field, competition will increase not decrease.

2. Customer service will be the key distinguishing factor. Soon products

and services will be commodities and customer service will differentiate the winners from the losers.

3. Customer service departments will turn away from software that thinks for reps and instead use higher quality personnel on the front line to make decisions.

4. Quality customer service labor will become more expensive. Anything that assists the front line reps without tieing their hands will not only improve sales but also the bottom line.

5. As corporate large scale consolidation continues front line reps will only have a limited knowledge of the solutions to problems and issues. As a result the problems and issues will have to be electronically handed over to subject matter experts.

6. Front line "call screeners" will be used more often.

7. Communication with the customer will come in various forms ranging from paper to electronic. Customer service reps will have to process and manage all forms.

8. Customers will soon be able to track the progress of their complaints through the Internet.

9. Satisfying the customer will become more difficult and customers will become more demanding.

— Oasis Technology

Companies are increasing their support functions by using software and hardware automation as productivity tools. Companies are recognizing that productivity tools cut costs for the company in the long term. There are a lot of underground costs that occur and can be decreased by software packages. Once companies become aware of the money they are wasting to get results, they will begin to invest in help desks.

— Fujitsu Software

Consolidation of multiple internal IT help desks

The main task of the traditional help desk is to provide a mechanism for "putting out fires." These help desks are designed to deal reactively when a server goes down, a printer malfunctions, or an office worker is unable to open a document. This is nothing more than applying a "quick fix" to users' immediate problems.

Most organizations follow a segmented approach to establishing help desk support. Different technology groups provide service for separate support domains, such as applications, LAN, WAN, PC hardware, PC software, etc.

When users have problems, they contact the appropriate help desk and the technical staff attempts to solve the problem.

When a problem arises in the distributed environment, though, it is likely to involve multiple systems, as well as the underlying networks. Diagnosing a problem requires the ability to analyze and assess all the elements on the critical path: the servers, workstations, network links and applications. The traditional help desk approach supporting separate technologies cannot respond effectively. The result is inefficient and ineffective support and end-user confusion about where to go with problems and which management group has responsibility for specific problems.

What is required is a broader operations and support function, based on either a centralized or distributed infrastructure, with end-to-end support capabilities. This new approach to support requires a complete revision of the help desk model and related support processes. Support of end-users cannot be separated from the use of IS tools for managing the distributed environment.

The new help desk model focuses on the needs of end-users for immediate, proactive support and service, as well as the tools and processes required to fulfill those support and service requirements. The shift requires bringing together support for end-users' problems with the planning, implementation, and management of the services they require. In other words, the new help desk integrates problem management with change, asset and service request processing.

Support organizations must establish enterprise-wide support processes using consistent and highly interoperable tools. This provides IS the ability to proactively manage and support the corporation's distributed computing assets. This approach is based on the concept that the help desk forms the foundation for implementing IS processes and establishing support competencies.

— Peregrine Systems

The future of help desks will be in their ability to transcend the traditional help desk role of a specialized tool for technical support departments into an enterprise-wide resource accessible to every worker equipped with a PC. This metamorphosis is already taking place. Help desk systems are being designed to coordinate a variety of key enterprise-wide management tasks and functions including: problem management and field service management as well as business processes and methodologies themselves. The best help desk system will be the one that can think for you, be easy to use and empower you with the ability to solve problems. The most intelligent software in the world will be the one that seems simplest.

— Professional Help Desk

In the next five years, the most significant development will be the centralization of the help desk as the brain trust of all information. Internet-driven data, network control, problem resolution, project, change and asset management will all be connected to and reside within the help desk infrastructure. This trend will be driven by cost-conscious managers who are looking to increase the timeliness and reliability of knowledge, and to ultimately reduce the total cost of ownership of IT assets across the enterprise.

— Magic Solutions

Web enablement, changing customer expectations, and wireless communication for remote service workers will bring about dramatic changes in the way that organizations support their customers. Although only a first step, empowering customers with the ability to log and check the status of problem and service calls via the Internet will eliminate waiting in queues for the next available support person as well as reduce duplicate or incorrectly logged service calls. Advantages will be that calls will close more quickly and vast knowledge bases will be created, thus raising customer satisfaction. Additionally the adoption of wireless technology will allow service technicians to receive, perform their work, and close calls without support personnel intervention.

— Metrix

- Continued movement to the Internet to provide additional customer support opportunities.
- Movement to consolidated help/service desk that provides single point of contact and supports all user environments.
- Customer support organization increasing in importance within the corporation.
- Shift away from "Level 1 and Level 2" style of support. End-users want to talk to talk experts on the first call.
- Standards will begin to emerge to allow heterogeneous help tools to work together. Help desk applications will become a commodity (all look alike).

— Allen Systems Group

The internal help desk will begin to more closely align with the function that interfaces with external customers. Obvious parallels exist between the customer support call center and the help desk itself but evolving similarities include:

1. Just as a support department dispatches a field service engineer (FSE)

to a customer site, an internal help desk will dispatch an analyst to support an employee;

2. An area often overlooked is the tracking of spare parts. A great deal of effort is spent tracking spare parts as part of the external support process. Interestingly, the same opportunity exists for tracking the internal assets supported by the internal help desk. This has the potential for a huge cost savings for a company;

3. Just as the support department logs defects and enhancement requests into a quality system, those IT help desks that do some form of development also have the need to capture defect/bug information and enhancement requests about applications that have been developed internally;

4. The use of Web-based trouble tickets that are dynamic by service level agreement. As the IT department negotiates the service level agreements with each business unit, this resulting agreement needs to be tracked and the future value comes by translating the terms into a dynamic environment. For example, the sales rep who logs a trouble ticket via the internal Web may see a Web page in one form whereas an engineer who logs a trouble based upon the SLA negotiated would see an entirely different trouble ticket. This ability to dynamically tailor the service is key. By providing the various employee segments (like segmenting the customer base), the help desk can provide highly focused service;

5. Lastly, the help desk will quickly evolve from IT to cover any form of organized employee support.

These are but a few of the emerging trends.

— Bob Tate, Vantive

Intel's Philosophy

Intel's PC OEMs were faced with increasing support costs with the continuing proliferation of calls from home PC users. As a result, Intel stepped into the picture to offer some well-researched advice.

"Our PC OEMs are getting killed with call center support costs," says Intel's Dr. Guy Blair. "They're spending five percent of their manufacturing costs on support."

In addition, Intel found their PC OEMs were spending too much time diagnosing and resolving problems. "Over 25% of their calls come in because of wrong software or drivers installed," says Blair. This, for example, is a type of problem that could be diagnosed and solved in minutes as opposed to a couple of hours, as it often does, explains Blair.

Blair says that in this situation, much of the time is spent reading the AUTOEXEC.BAT file. But if the support rep can use remote diagnostics (taking over the end-user's PC) they could simply reboot the machine and close the ticket in three or four minutes.

At industry trade shows, Intel has been demonstrating their "proof of concept" demo which incorporates five technologies they believe PC OEMs could use to improve remote diagnosis and to install software upgrades remotely to cut costs and save time.

While Intel does not make the technologies they advocate (they sell chips to their PC OEMs), they talk to their partners and test such technologies in their Architecture Lab. It's in their best interest to do so. "We are enjoying watching the explosion (of PC sales) and selling more chips so we want to make sure our PC OEMs are not running into a wall when it comes to support," says Blair.

Here are the technologies and their descriptions:

TAPI - Telephone Application Programming Interface, also called Microsoft Telephony API.

WinSock 2 - A Microsoft operating system standard that supports multiple transports: modem, ISDN, LAN, ATM, allowing independent applications to use the same connection.

DSVD - Digital Simultaneous Voice and Data. This is a standard for transmitting voice and data at the same time over a standard analog phone line. Using a DSVD-capable modem you could talk to someone while data is being transferred on the same phone line.

DMI - The Desktop Management Interface. As described by Intel, DMI is the result of a cooperative industry-wide effort to make PC systems easier to manage, use and control. DMI describes hardware and software components in a format that can easily be accessed by PC management applications and technical support reps.

This would allow them to view inventory of hardware and software components, view and change parameter values and settings and view data generated by software agents and diagnostic routines. DMI was developed by the Desktop Management Task Force (DMTF) which is a consortium of hardware, software and peripheral vendors.

Remote control/application sharing - There are many products that make this possible, letting users take over another PC from their own PCs so they can diagnose and fix a problem remotely.

The Intel Architecture Lab's research identified three ways to resolve the call center support dilemma: call avoidance, call deflection and call handling. The tools named above are being used in two of their call centers with a third one to follow.

Other technologies they see as reducing the need for technical support calls (as part of their three strategies) include connecting to the Internet to interact with an expert system, using wizards and expert systems to detect and resolve PC problems, using software agents that can monitor the PC to detect and automatically resolve several types of PC problems, and using VRUs and ACDs.

"We want our PC OEMs to know about these technologies, independent of how they evolved, so they can be ready for the future," says Intel's Dr. Jim Larson.

He says to come up with these ideas, they talked to their PC OEMs, tried

Smooth Sailing Tips

The help desk should be able to handle any number of end-users (preferably simultaneously).

Help desk software should have the ability to store knowledge about the intricacies of any type of hardware and any type of software.

Help Desk software should he able to distinguish between urgent requests for help (i.e. priority for the mobile sales force) and other types of requests.

All questions, whether they be from naive end-users or power-users, should be able to be answered from the help desk's knowledge base.

The Help Desk software used should be able to make "experts" out of junior level help desk analysts.

Ultimately, the help desk should act as an extra staff of support personnel. To do so requires that the software have the functionality - and the intelligence - to tailor itself to the organization and grow with it as the organization grows.

— Professional Help Desk

their ideas out in their own internal call centers and did some prototyping.

While there are existing modems that incorporate DSVD, they are not recognized by Intel for DSVD because Intel says that as the technology now stands it's proprietary, meaning that there are no guarantees that any DSVD modem can talk to another.

The benefits in using a DSVD modem in conjunction with remote control/application sharing is that it lets support reps talk to the end-user while accessing their PC at the same time, makes it faster and easier to diagnose and make changes to user PCs. There's no need for either party to hang up and wait for the modem connection to transfer data.

Blair and Larson say Radish Communication Systems (Boulder, CO) is closest to delivering a real product that will incorporate DSVD, DMI and remote control/application sharing. Blair says Microsoft and Artisoft/Triton are also working to implement the five technologies Intel has been evangelizing.

Radish's VoiceView has been adopted by major PC OEMs like NEC, Packard Bell, Hewlett Packard and AST, along with major modem manufacturers as an industry standard communications protocol. Radish's software, TalkShop, is a popular application for technical support, requiring only a PC, a VoiceView-enabled modem and analog or digital phone.

In a support application, PC or modem manufacturers can provide an interactive support button in TalkShop so when a customer needs help they can click this button to reach a support rep on the phone. The TSR can diagnose and fix problems by remotely modifying files and executing commands

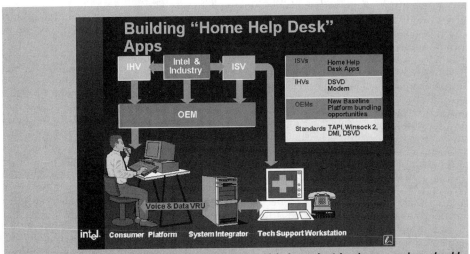

It's Intel's opinion that the independent software and independent hardware vendors should work to deliver standards in the technologies they see changing support in the future for their PC OEMs.

on the customer's PC while on the phone with the customer.

Using VoiceView's TalkShop 2.0, OEMs can choose to ship their PCs with modems and software that support this new release. Some of the enhancements will be new remote control features; a DMI browser so support reps can view a TalkShop client's DMI configuration (with permission); and SVD support (which they say is the DSVD specification endorsed by vendors like Hayes, Rockwell, Creative and Intel).

The VoiceView software, in the simplest application, lets two people with VoiceView enabled modems discuss a set of files over the phone, then switch to data and transfer the files at a full 28.8 Kbps and then switch back to voice mode to talk some more — all in one phone call. Thus, for a period of time (when the data is transferring) conversation cannot take place.

"It's not that we haven't had the technology to develop simultaneous voice and data, it's just that until now it was too expensive to support DSVD, says Radish's Paul Davoust. "Now the price has come down, but there are still some costs associated, in addition to the simultaneous mode requiring more horsepower."

Davoust is questioning whether or not the cost will still be an issue. "Most will think that it's great, but even with prices down it's not free, so will they (the modem manufacturers) put it in," questions Davoust.

Intel thinks that their PC OEMs should start to demand some of the technologies they suggest, because only then will they begin to see support improve and associated costs for support drop.

It may be a little while before you actually see the technologies Intel is advocating actually put to use. While there is little doubt that these technologies can drastically improve support and cut support center costs, some of the standards and technologies are not quite there yet.

The smaller support centers should keep a watchful eye on Intel's large PC OEMs to see if and when they start to adopt these technologies as they become available or further developed.

For more information, Intel has written three White Papers on the technology. You can find them at Intel's Web site, Intel.com.

Support Standard Initiatives

New standards soon to be introduced will change the face of support — making it less costly, and giving you more choices and fewer integration headaches.

The Desktop Management Task Force is busy working on developing two new standards for the support industry: an exchange standard for passing electronic service tickets between companies; and an import/export standard for support knowledge, designed to make knowledge more widely available.

Often, after a support rep creates a ticket based on the customer's problem, he or she will realize that problem ticket needs to be passed to another vendor or a different support organization within the company. When this happens, the process of creating a ticket has to be repeated each time the ticket is passed, often back and forth several times.

When external PCs are not networked and different organizations use different management systems, the ticket needs to be re-read and re-entered — wasting valuable time. Sometimes information gets lost. This new exchange mechanism (expected to become a standard soon — perhaps by the time you read this) will surely improve the efficiency and effectiveness of support when more than one organization is involved in problem solving.

It will also eliminate some of the calls that are required to repeatedly describe or update the same ticket. The standard will make it possible for anyone with multi-vendor systems to share information, without having to integrate their systems.

"There needs to be a common definition for passing tickets," says Patrick Bultema of the Bultema Group, who is chairing the Support Management Working Committee formed by the Desktop Management Task Force (DMTF).

"We're working to identify import and export contents and agree on a format. It's really just a simple data standard." Bultema is the former chairman of the Help Desk Institute.

DMTF is a "cooperative industry-wide effort to develop and deliver the enabling technology for building a new generation of PC systems and products." It is led by a steering committee that includes Compaq Computer, Dell Computer, DEC, Hewlett-Packard, IBM, Intel, Microsoft, NEC, Novell, Santa Cruz Operation, SunSoft and Symantec. Most recently, DMTF deliv-

Smooth Sailing Tips

Ask your technical support staff for the most common problems they receive questions about. Build an expert system based on those problems and put it on your web site. There is guaranteed payback. You know these problems are frequent, and automating the answers will free support staff to work on other problems, and make for more satisfied customers. Since they are commonly occurring problems, your technical support staff knows the symptoms and how they are resolved; making system development very easy.

Du Pont, a world leader in the use of expert systems, recently reached $1,000,000,000 in savings, due to their wide-scale implementation of these systems. Du Pont provides all 130,000 employees and subsidiaries with access to easy to use expert system tools.

— Exsys

ered the industry standard Desktop Management Interface (DMI).

To create these two new standards DMTF formed the Support Management Working Committee whose chart members include the following vendors and consultants: Advantage kbs, Applix, Astea International, Bendata, The Bultema Company, The Bentley Company, Clarify, Digital Equipment, Folio, Foresight Software, Hewlett-Packard, High-Tech/High Touch, Hyperion Associates, IBM, Inference, Intel, KnowledgeBroker, Keane, Magic Solutions, Peregrine Systems, Professional Help Desk, Quintus, Remedy, Software Artistry, Scopus, Software Support, Vantive, Ziff-Davis/Support On Site.

Bultema says both the electronic service ticket standard and the import/export standard for support knowledge will be wrapped up and released by January 1997.

High Exceptions for Standards

The demand for prepackaged knowledge has, in many cases, not been adequately filled. Some large companies have built their own knowledge bases. But the time and expense of hiring knowledge and case writers who they must keep on staff to make changes and continue to add new knowledge is daunting.

Future Trends

Here's a look at what else will change as we look into the future of support. Some of these tips come from Pat Bultema's forum at Support Services Expo in March of 1996 on Future Trends.

Alternate modes of delivery. Bultema predicts that soon, about half of the volume of service requests will come by means other than the telephone (i.e. Internet and e-mail requests).

You should be able to manage all service requests through the ACD queue, not just the "calls" that come in. But you can't, not yet. Bultema says tools and technologies need to be developed to manage these "multiple queues."

You will be able to run products out of the box on all levels. You'll even be able to install the high end systems out of the box, and systems will run on less expensive hardware.

Bultema says you'll see new support architecture that creates plug and play. There will also be more middleware. Bultema suggests avoiding high priced labor intensive products such as UNIX boxes, where the price of the box ends up outweighing the cost of the software.

Bultema says there are currently 188 problem management vendors (in 1996), fewer than last year. Several products from some of these vendors are really taking off, generating a sub revenue and a consolidation of market share.

This will drive volume and put downward pressure on price. The low volume companies won't be able to survive. You'll also continue to see more mergers and acquisitions. When shopping, it's a good idea to use the company's stock performance as indication of the company's strength.

Some help desks have chosen to buy bundles of prepackaged knowledge from outside companies that sell problems and solutions to common software and hardware issues.

DMTF believes that there is not a lack of useful knowledge or knowledge systems, but that the problem is a lack of a consistent method for capturing and porting knowledge into a knowledge tool the help desk already uses.

With this standard, knowledge publishers will be able to create a wider range of more useful knowledge in a universally accepted and understood platform. It will make it possible for you to buy any published knowledge you need without worrying about how to import it into your existing systems.

The committee is currently working to find an agreed upon content, structure and exchange format for knowledge, something not technically difficult.

As described in the DMTF's white paper, the content of a case in the standard is expected to include these four elements that make up a case:

- symptom information that describes the problem or condition that necessitates support;
- the relevant elements of environmental information required to identify a particular problem or condition;
- the diagnosis of the problem;
- the action required to solve the problem.

The standards-body also needs to agree on the structural terminology, mainly the field names for the fields in a case. They are also considering a way to group related cases together.

In addition, the standard will define a format used for passing knowledge, likely one of the currently accepted data formats.

Bultema says that instead of only being able to choose from the approximately 700 case base bundles that are available for purchase now, this standard will bring about a choice of about 140,000 different case bases in the future. "It will be easy to create specialized applications," says Bultema. There will be less work to do."

He says help desks will be able to simply download these blocks of knowledge. He also says the standard will especially help niche markets. For example you may soon see AS/400 trouble packs that would not have been created because the demand was not great enough.

Bultema says it's taken this long to develop such standards because of the industry's immaturity. "They've been more focused on building business rather than standards, but [the industry] seems to have reached its pain threshold."

What's surprising is that, according to Bultema, the fierce competitiveness that is rampant among the help desk vendors has taken a back seat during negotiations for developing the standard. "In the past standards failed because of vendor control. But now there is a consensus," he says. "They're all cooperating."

By the end of 1996, an initial document will be passed along to all of the charter members who will make suggestions for refinement and by January 1997 an initial version is expected to be released.

While these standards will not solve all of your support nightmares, it's a start in the right direction. Bultema says that once these two standards are passed, the committee will continue to work on new projects. The road into the future of support should have fewer bumps and dead ends.

Multi-vendor Support

Problems that seem easy to solve at first can get very frustrating when the customer's real problem lies with another vendor's product. The obstacles associated with providing multi-vendor support may soon be a thing of the past.

When someone has a problem with their PC, their first instinct would probably be to call the manufacturer. Yet the computer is made by HP, the software loaded comes from Lotus, Microsoft, WordPerfect, the internal e-mail system comes from Novell and all PCs are networked on IBM's LAN Server. In reality, the problem could be occurring with any one of these systems.

The transition from working in standalone mainframe environments to networked PCs that share data and applications has made technical support more complicated and costly. Working in this new distributed environment means companies can face problems with one of several platforms or applications.

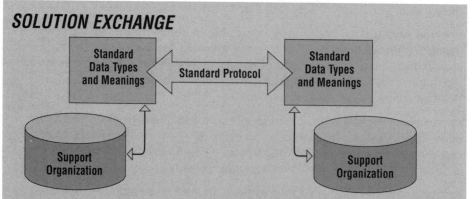

SOLUTION EXCHANGE

The Solution Exchange Standard provides a way for companies to exchange solutions and solution information without losing any of the content or value in the process. (Source: Customer

Who's the right person to call? The answer is: who knows? All the end-user knows is that they need to be up and running right away. Time is money. They don't care who is responsible for fixing the problem, as long as it get fixed and they don't get the runaround.

The customer expects that the vendor they dialed first is responsible for fixing their problem. To tell the customer to call another vendor and start from scratch is not acceptable. The customer is irate at having to repeat the problem two or three times.

This forces the first vendor to work with other vendors to solve the mutual customer's problem.

It may also be beneficial for other reasons. For example, if someone is having a problem getting hooked up to the Internet, they may call their online carrier. The online carrier may realize that the customer is having a problem with their modem, and therefore, the responsibility to fix the problem is not theirs.

But the sooner the customer's modem is working, the sooner they can get online, spending money. It's in the best interest of the Internet service provider to contact the modem manufacturer to try and solve the problem. Telling the customer they should call their modem manufacturer may mean the customer will put off the task. Or the modem manufacturer may be hard to reach for support and not be available all of the time.

End-users want to be able to call one number for any support issue, but it's just not that easy. For one thing, some of the problems get complex, requir-

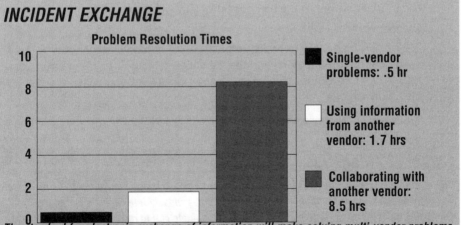

INCIDENT EXCHANGE

Problem Resolution Times

- Single-vendor problems: .5 hr
- Using information from another vendor: 1.7 hrs
- Collaborating with another vendor: 8.5 hrs

The standard for electronic exchange of information will make solving multi-vendor problems solving much easier and faster.

Solving a customer problem takes more than three times as long when the resolution involves another vendor's product — if you can use readily available information. But when you can't use information from another vendor, problem resolution takes 17 times as long as a single vendor problem. (Source: CSC.)

ing technicians from competing companies to put their heads together to resolve the problem. This is very time consuming and costly.

But an even bigger problem with multi-vendor problem solving is that there is no way for one vendor to pass a trouble ticket to another vendor. The problem must be repeated or re-entered. Even though these multi-vendor support issues are not all that common, when they do occur, they take a very long time to solve.

Realizing how important this issue is, top software and hardware manufacturers have formed committees with trade associations to come up with solutions.

Like The Desktop Management Task Force (working developing an electronic exchange mechanism, The Customer Support Consortium (Seattle, WA) has also been taking steps to address the problems associated with solving multi-vendor problems. The CSC and DTMF have teamed up to avoid duplicating efforts and to ensure that only one standard is developed.

The CSC advisory board consists of representatives from companies like Novell, Lotus, Microsoft and Dell. Based on their analysis of the industry, CSC and their members came up with what they've termed Solution-Centered Support.

This idea of Solution-Centered Support is based on creating a protocol where companies who provide support to their customers could transfer both solutions and open problem tickets or product information to another company regardless of the call tracking/problem resolution tools being used.

Being able to electronically pass this information would greatly reduce the time spent solving a multi-vendor problem when two or more vendors are forced to collaborate.

"It's one standard with two elements — the solution exchange [for knowledge capture and export] and the incident exchange [for transporting open calls from one vendor to another]," says Greg Oxton, executive director of the Consortium.

This model recognizes that many multi-vendor problems can be solved

Smooth Sailing Tips

One of the most important things support managers can do to make their operations run more smoothly is use support technologies that integrate seamlessly into their analysts' workflow. For instance, if significant time must be spent after a phone call to record the solution the analyst provided, time that could be spent helping other customers is being wasted. Analysts should not be made to spend time recording and tracking the knowledge they produce - their job is solving customer problems. The capturing and sharing of support solutions should be performed automatically by the problem resolution system, as a dynamic function of the problem-solving process.

— Primus

through the use of information about another vendor's product, and offers a set of standard ways to share information and a Solution Exchange Standard.

Since their model takes into consideration that some multi-vendor problems can only be solved by two organizations working together, it offers a set of standard methods.

The multi-vendor model is based on tiers. Tier I is a way to share solutions among organizations that solve multi-vendor problems. A retrieval system will accept a query and route it to multiple vendors for their information. It will then return the information to the first vendor's Internet browser. The support tech will have access to information from all participants with just a mouse click. This will also give companies a standard channel for publishing information.

Tier II deals with customer problems that cannot be solved with Tier I information alone. It provides a channel for collaboration with another organization as more of a business process.

Tier III provides standards between companies (mainly vendors) that need to work together on frequent or extremely challenging problems. Oxton says Novell and Microsoft are already piloting a working model of Tier III.

Each support organization can decide which Tier to adopt based on the relationship they want to have with the support organizations they need to work with.

The standard will also make it possible for a problem to be solved once and shared between all vendors. By the time you read this article, a working model of the standard should exist, according to Oxton.

The CSC has adopted Scopus Technology's (Emeryville, CA) DDTP (Digital Document Transfer Protocol) as the standard protocol for use in exchanging electronic documents between companies with different computer systems.

According to Carter Lusher, VP and research director, Management of Technology, The Gartner Group, a working model should exist by the time you've read this book. But the real question lies with the tool providers, and whether or the not actual help desk/customer information system vendors will accept this working model.

Lusher says it could go either way. "There is a hesitation there, a wonder if this will commoditize the marketplace or help the marketplace to grow with ability to put together a best of breed tool set," says Lusher.

Still, Scopus has a safety net in mind. Oxton and Bahram Nour-Omid, Scopus's chief technology officer, say that companies like Novell (who is using Vantive's help desk product) and Microsoft (using Clarify's system) will put pressure on such vendors to make the protocol a part of their product for easy ticket exchange and knowledge transfer.

"If they still don't accept it, we will work with the end-user groups (like Microsoft and Novell) to make the exchange work," says Nour-Omid.

"Scopus has been very aggressive in investing a great deal of time and energy in developing this protocol," says Carter. Nour-Omid says it also makes their product more valuable and appealing for companies looking at "enterprise workflow."

"We realized the need for messaging systems to exchange information without too much trouble in a way that everyone can adapt to," says Nour-Omid.

The three constituencies that stand to gain the most with the new standard are the technology sellers (like Novell and Microsoft), the outsourcers and those who contact the help desk.

Another thing that must happen: the Desktop Management Task Force must approve the protocol. Lusher says conversations between the two groups are in the works but we will have to wait and see what happens.

"It's our opinion that the two groups work together, merging as one or complementary, dividing responsibilities," says Lusher. He says the CSC's strength is in the fact that they've been working on this for a couple of years with interest from their members and have been testing their ideas. But they are still a Consortium. It's the DTMF that can actually ratify and implement the new standard.

"The work could be accelerated if both groups work together but there may be different missions or agendas, as in any two groups," says Carter.

The first quarter of 1997 is the earliest we will see a standard ratified and accepted, says Carter. Then by mid-year 1997 we could see products that actually comply.

Streamlining E-mail Support

E-mail has become a popular option for requesting support, but not without some drawbacks. A new service aims to take away the drawbacks of "E-mail tag" saving support centers and customers time and money.

When end-users with technical problems send hardware or software vendors an e-mail message it often requires at least two more e-mail transactions before the vendor can solve the problem. This is often overly time-consuming and frustrating, barely more desirable to the customer than a 20 minute wait on hold for live support.

No doubt many of the Web tools from the help desk vendors make it possible to let end-users try and solve a problem on their own. Often when the system uses some kind of intelligence, there is a reasonable success rate. But the customers who are not successful must call for support or send a general e-mail to the help desk describing their problem.

Typically the end-user does not include enough information in their e-mail message for a support rep to solve the problem. Typically lacking will be PC configuration information or other significant details the rep needs to know about the problem experience.

What if all e-mail could come in a standard format with the right diagnostic information to enable the rep to solve the problem right away and e-mail back a solution (along with any software files the customer may need)? Now that can actually happen.

The technology is called e.support from Touchtone Software (Huntington Beach, CA, 714-969-7746) and The Software Support Professionals Association (SSPA, San Diego, CA, 619-674-4864). The e.support system guides online users through fill-in support request forms, which generate individual problem reports. A system diagnostic file is automatically attached to the request before it's sent to the vendor the end-user is trying to reach for support.

E.support was developed by Touchtone at The SSPA's request. The SSPA is heading the effort in response to what they say their members need. They partnered with Touchstone, because Touchstone had already started working on the technology to make e.support possible.

Here's how it will work:

Vendors who would like their customers to use this service, sign up with Touchtone who acts as a clearing house, charging the vendors who incorporate e.support a fee. Touchtone then sends the vendor the software they need to read diagnostic information and send files, and the software their customers need to send the e-mail.

By the time you read this, the software should be available on the Internet free of charge to anyone who wants to use it to send e-mail to vendors they require support from. If these vendors are not signed up in the e.support program, they will still receive the formatted e-mail, minus the diagnostic information. Users who do not have an Internet account require only a modem.

Smooth Sailing Tips

Make sure that all of your business processes are integrated into your technology solution.

Integrate your telephony system into your help desk technology as soon as possible.

Ensure that you have SLA's in place for all of your customer groups. Integrate your SLA's into your help desk technology.

Understand your customers on a more personal level. Meet their individual service needs.

Translate your service performance into a dollars and cents impact to your customer's bottom line.

— Amdahl

The system can still benefit companies that are using Web tools from their help desk software vendors to allow customers to seek support through the Internet. "Often, home market users who use these tools can't form a query or figure out their problem," says Touchstone's Carolyn Gross. "e.support will complement these tools."

Gross says Touchstone is creating an API so users can integrate the e.support software with their help desk/call tracking system, so a ticket can be generated. As the system exists now, the customer must decide who to send the e-mail to, whether or not they are sure who is responsible for fixing their problem. But Gross says they are also working on a feature that will add intelligence, re-routing the e-mail to the vendor who is responsible for the problem even if that is not clear to the end-user. Another feature to be added will automatically reply to commonly asked questions.

Touchstone also plans to release a version of e.support for the internal help desk that supports corporate employees.

While Touchstone hopes most vendors will agree to a standard 24 hour response time for free support when responding to e.support e-mail, they cannot guarantee this. Response rates to e-mail may vary from vendor to vendor. However, Gross says most should commit and the SSPA can put some pressure on their members. In some cases, vendors may offer to respond to customers in two to six hours if these customers are willing to pay for priority service.

It's hard to question the value of e-mail for the help desk. Most vendors are likely to see e.support as a way to provide efficient support at no cost to their customers and with minimal effort on the part of both themselves and their customers.

Dataquest estimates that the current cost for companies that let customers use the Internet to request support or try to find solutions to their own problems, is $2 to $3. That's a big savings over handling support calls. Dataquest's Tom Sweeny says that to handle every call live costs an average of $20 to $25 depending on the complexity of the problem.

Still, using e-mail for support has some drawbacks. For one, if a user's PC is down, they cannot send an e-mail request for support.

Secondly, sending an e-mail is not as comforting as telling your problem to a sympathetic human ear whose job is to offer you a solution. Using e-mail takes away the human interaction element.

Yet when customers are certain they can save many precocious minutes posting a problem online for a speedy response as opposed to holding on the phone, they start to see e-mail as a viable option.

Broderbund Software is beta testing the e.support system. So far, Broderbund's Jim Wilmott says they've only been involved in limited testing and no agreements have been reached yet, but they are excited about the prospects.

"Our customers want e-mail support," says Wilmott. Currently Broderbund uses Inference's help desk system for solving problems over the phone and for letting customers use Inference's Case Base Reasoning engine to solve their own problems.

"We've had enormous success using Inference for Web support," says Wilmott. "And it's user-friendly. But some of our customers who have never encountered such a tool use the general e-mail option. With the diagnostic features of e.support we can eliminate multiple contacts [through e-mail] with the customer. It's like looking at the customer's system as opposed to pulling the information from the customer."

To give you an example of the tremendous growth of customers using e-mail for support, Wilmott says Broderbund did not have a Web page until November of 1995. But as early as January of 1996, they received 1,600 requests for support though e-mail.

"We're also beginning to ship the AT&T Browser to our customers and believe that this will have a direct effect on the number of customers who go though the Internet to reach us for support," says Wilmott. The AT&T Browser will offer customers who install the software a certain number of free hours on the Internet each month.

One fear Wilmott says they have with using e.support is that customers may over use the service. "It's the same fear as having an 800 number,"

Tips for Buying Problem Resolution Tools

• Be sure you have gotten the most out of your existing investment.

• Be sure you understand your information management requirements.

• Be sure you follow-up with references in detail.

• Be sure you can integrate your new technology purchase into your existing technology infrastructure.

• Establish a return on investment business case so that you can measure the success of your investment.

• Beware of the differences between key word search, decision tree, case-based reasoning, and heuristic reasoning as far as technologies for problem resolution go. Different vendors use different technologies and some will be more appropriate than others for your business.

• Understand that problem resolution is different than call tracking and problem management and some vendors only sell one of the layers.

• Buy a technology solution that is scalable and start with the size that is appropriate for your current operation.

— Amdahl

says Wilmott. "Customers may overflow the system when there is no fee." Wilmott says the biggest fee their customers ever pay for support is the price of a phone call. Everyone who buys their software is offered free support.

By the end of the year Touchstone expects to have a clearing house where all support requests will be forwarded to several thousand vendors; they expect to introduce full Internet connectivity; a feature that will automatically reply to commonly-asked questions; and some additional features.☾

Closing Thoughts

How these changes will affect you

All this talk of changes and improvements coming down the line are sure to help make support more cost-effective and less complex. The help desk industry is still a very young industry. Many of the leading technology vendors have been around for a mere five years or less.

Think back to when you were ten-years-old. Do you remember people using computers, except perhaps government agencies or motor vehicle offices? Unless you are under 30, you will probably be pressed for a memory.

Twenty-five years from now, today's "Generation X" are likely to be astounded at how antiquated the business world used to be. I find it hard to believe my parents lived without television until it was invented when they were teenagers. I'm sure that in 25 years, a new generation of people will find it hard to believe that before the 21st century people used to type on keyboards instead of simply speaking to their computers.

But even with the drastic changes you're sure to witness in years to come, the strategies and philosophies put into place in your help desk today will carry you into the future. Work incentives, fair treatment of employees, a clear vision, being goal-oriented and the continued philosophy that the customer is the lifeblood of your business, will not become outdated, regardless of the changing technology. These strategies are what will separate the model help desks from the unsuccessful help desks.☾